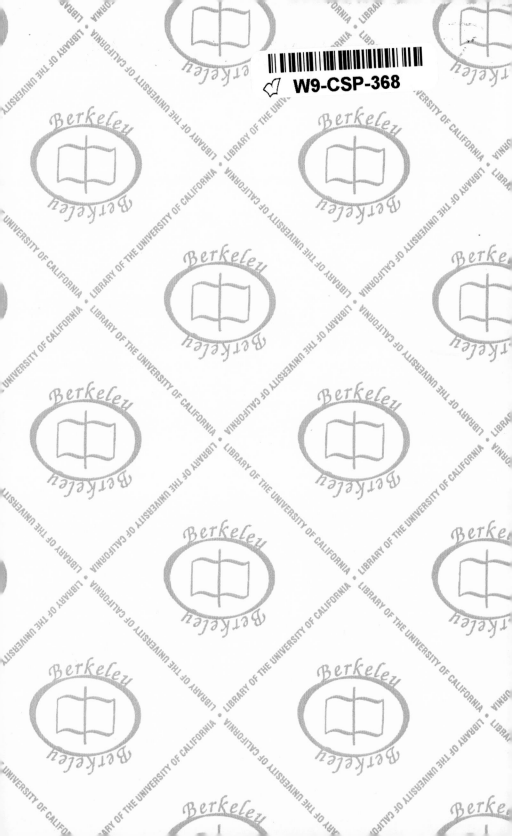

CAMBRIDGE STUDIES IN MODERN OPTICS: 2

General Editors
P. L. KNIGHT
Optics Section, Imperial College of Science and Technology
S. D. SMITH, FRS
Department of Physics, Heriot-Watt University

Optical holography

Optical holography

Principles, techniques and applications

P. HARIHARAN

CSIRO Division of Applied Physics, Sydney, Australia

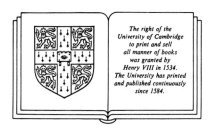

The right of the
University of Cambridge
to print and sell
all manner of books
was granted by
Henry VIII in 1534.
The University has printed
and published continuously
since 1584.

CAMBRIDGE UNIVERSITY PRESS

Cambridge

New York   Port Chester   Melbourne   Sydney

Published by the Press Syndicate of the University of Cambridge
The Pitt Building, Trumpington Street, Cambridge CB2 1RP
40 West 20th Street, New York, NY 10011–4211, USA
10 Stamford Road, Oakleigh, Melbourne 3166, Australia

First published 1984
Reprinted 1985
First paperback edition 1986
Reprinted 1987, 1989, 1991

Printed in Great Britain at the University Press, Cambridge

Library of Congress catalogue card number: 83–15063

British Library cataloguing in publication data
Hariharan, P.
Optical holography – (Cambridge studies in modern optics)
1. Holography
I. Title
774   QC449

ISBN 0 521 24348 3 hardback
ISBN 0 521 31163 2 paperback

MP

To Rajeswari

Contents

Preface

The last ten years have seen an upsurge of interest in optical holography because of several major advances in its technology. Holography is now firmly established as a display medium as well as a tool for scientific and engineering studies, and it has found a remarkably wide range of applications for which it is uniquely suited.

My aim in writing this book is to present a self-contained treatment of the principles, techniques and applications of optical holography, with particular emphasis on recent developments. After a brief historical introduction, three chapters outline the theory of holographic imaging, the characteristics of the reconstructed image and the different types of holograms. Five chapters then deal with the practical aspects of holography – optical systems, light sources and recording media – as well as the production of holograms for displays and colour holography. The next two chapters discuss computer-generated holograms and some specialized techniques such as polarization recording, holography with incoherent light and hologram copying. These are followed by four chapters describing the more important applications of holography. Particle-size analysis, high-resolution imaging, multiple imaging, holographic optical elements and information storage and processing are covered in two of these, while the other two are devoted to holographic interferometry and its use in stress analysis, vibration studies and contouring.

To make the best use of the available space, the scope of the book has been limited to optical holography. No attempt has been made to cover related techniques such as acoustical and microwave holography, which have not made such rapid progress. In addition, much of the material on basic concepts of optics found in earlier books on holography now forms part of many introductory science and engineering courses. This has therefore been summarized in a set of appendices, where it is available for reference.

This book is intended for people who would like to learn more about optical holography as well as those who would like to use it. Students will find the book useful as a supplementary text, while researchers can use it as a reference work. The initial presentation of each topic is at a level which is accessible to anyone with a working knowledge of physical optics. This is then followed by a more detailed treatment for the serious worker. References to about 700 selected original papers identify sources of additional information and will hopefully guide the reader through the voluminous literature in this field.

I am grateful to many of my colleagues for their assistance. A special debt is due to Dr W. H. Steel who has been a continuous source of encouragement and helpful criticism.

Sydney, Australia   P. Hariharan
January 1983

1

Introduction

When confronted with a hologram for the first time, most people react with disbelief. They look through an almost clear piece of film to see what looks like a solid object floating in space. Sometimes, they even reach out to touch it and find their fingers meet only thin air.

A hologram is recorded on a flat surface, but produces a three-dimensional image. In addition, making a hologram does not involve recording an image in the conventional sense. To resolve these apparent contradictions and understand how a hologram works, it is necessary to start from first principles.

1.1. The concept of holographic imaging

In all conventional recording techniques such as photography, a flat picture of a three-dimensional scene is recorded on a light-sensitive surface. This can be done by a lens or, even more simply, by a pinhole in an opaque screen. What is recorded is merely the intensity distribution in the original scene. As a result, all information on the relative phases of the light waves from different points or, in other words, information about the relative optical paths to different parts of the object, is lost.

The unique characteristic of holography is the idea of recording the complete wave field, that is to say, both the amplitude and the phase of the light waves scattered by the object. Since all recording media respond only to the intensity, it is necessary to convert the phase information into variations of intensity. This is done by using coherent illumination, as shown in fig. 1.1, and adding a reference plane or spherical wave to the wave scattered by the object.

Without going into the detailed theory, it is apparent that what is recorded on the photographic plate is the interference pattern due to the two waves. The intensity at any point in this pattern depends on the phase as well as the amplitude of the original object wave. Accordingly, the processed photographic plate, which is called a hologram (or whole record), contains information on both the amplitude and the phase of the object wave. However, since the hologram bears no resemblance to the object, this information is in a coded form.

The reason for the success of holography is that the object wave can be regenerated from the hologram merely by illuminating it once again with the reference wave as shown in fig. 1.2. To an observer, this reconstructed wave is indistinguishable from the original object wave; he sees a three-dimensional image which, as shown in figs. 1.3 and 1.4, exhibits all the normal effects of perspective and depth of focus which the object would exhibit, if it were still there.

1.2.   Early studies

The roots of holographic imaging can be traced back to work by Wolfke [1920] and by Bragg [1939, 1942] in x-ray crystallography, which led to the development of the 'Bragg x-ray microscope'. When a crystal is illuminated with a beam of x-rays, the diffraction pattern obtained

Fig. 1.1. Recording a hologram. The photographic plate records the interference pattern produced by the light waves scattered from the object and a reference wave reflected to it by the mirror.

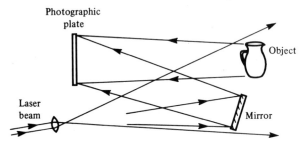

Fig. 1.2. Reconstruction of the image. The hologram, after processing, is illuminated with the reference wave from the laser. Light diffracted by the hologram appears to come from the original object.

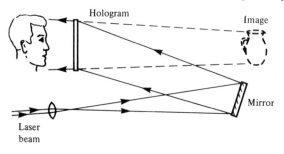

Fig. 1.3. Views from different angles of the image reconstructed by a hologram, showing changes in perspective.

corresponds to the square of the modulus of the Fourier transform of the amplitude scattered by the crystal lattice. As a result all information on the phase is lost. This is a major problem in the determination of crystal structures. However, for a centrosymmetric unit cell, the Fourier transform is real, and only the sign of the diffracted amplitude can change. In fact, for a crystal with such a structure, and with a heavy atom at the centre, the scattered amplitude from the heavy atom can be large enough that the weaker scattered amplitudes from the other atoms can add to it, or subtract from it, without altering its sign. If then, a mask is made whose transmittance at any point is proportional to the measured amplitudes in the x-ray diffraction pattern and illuminated with a collimated beam of monochromatic light, the Fraunhofer diffraction pattern obtained, which corresponds to a second Fourier transform, yields directly an image of the structure of the crystal.

Gabor's aim when he proposed the idea of holographic imaging [Gabor, 1948, 1949, 1951] was to obtain increased resolution in electron microscopy. Since it was difficult to correct the spherical aberration of magnetic

Fig. 1.4. Picture of the reconstructed image taken with the camera lens at full aperture (f/1.8), showing the effect of limited depth of focus.

electron lenses, he proposed to record the scattered field of the object when it was illuminated with electrons and then reconstruct the image from this record using visible light. To demonstrate the feasibility of his proposal, he used a setup in which light waves were utilized both to record the hologram and to reconstruct the image from it.

In Gabor's experiment, the object was a transparency consisting of a clear background with a few fine opaque lines on it. This was illuminated with a parallel beam of monochromatic light and its Fresnel diffraction pattern was recorded on a photographic plate. The complex amplitude in this diffraction pattern could be considered as the sum of two terms, a constant term due to the directly transmitted beam, which constituted the reference wave, and a varying term due to the light scattered by the fine details in the object. Because the reference wave was much stronger than the scattered wave, variations in the phase of the scattered wave resulted mainly in variations in the amplitude of the diffraction pattern, its phase being always very nearly that of the reference wave.

The actual hologram was a positive transparency made from this negative, the exposure and processing conditions being adjusted so that the amplitude transmittance of the hologram at any point was proportional to the intensity in the original diffraction pattern. When this hologram was illuminated once again with a collimated beam of monochromatic light, the transmitted wave had a uniform phase but exhibited the same variations in amplitude as those in the diffraction pattern. Because of these spatial variations in transmission, the hologram produced two diffracted waves, one corresponding to the original scattered wave from the object, and another with the same amplitude but opposite phase which formed another image (termed the conjugate image).

The similarities of Gabor's experiment to Bragg's x-ray microscope are evident, but the differences are also extremely significant. In the latter, because of the symmetry of the object there is no phase information to be lost and an exact reconstructed image can be obtained; in the former a much wider range of objects can be handled, but the loss of phase information, though tolerable, leads to the formation of an additional conjugate image.

Following Gabor's work, several attempts were made to produce holograms with an electron microscope. However, the successful application of his technique to this field has not materialized so far because of several practical problems.

Optical holography was also not very successful initially, even though the validity of Gabor's ideas was confirmed by a number of workers, and some of the later developments in holography were anticipated by them [Rogers, 1952; El-Sum & Kirkpatrick, 1952; Lohmann, 1956]. The main reason for the lack of progress was the poor quality of holographic images. This was largely due to the presence of the conjugate image as well as scattered light from the direct beam, both of which were superposed on the reconstructed image. Several techniques were proposed to eliminate the conjugate image but none was really successful. As a result, interest in optical holography declined after a few years.

The breakthrough which effectively solved the twin-image problem and opened the way to the large scale development of optical holography was the off-axis reference beam technique developed by Leith & Upatnieks [1962, 1963]. This arose out of their earlier work on the optical processing of synthetic-aperture radar data. They argued that the conjugate image was essentially due to aliasing, and introduced a spatial carrier frequency by using a separate reference wave which was incident on the photographic plate at an appreciable angle with respect to the object wave. Such a hologram, when illuminated with the original reference beam, produced a pair of images which were separated by a large enough angle from the directly transmitted beam and from each other to ensure that they did not overlap.

This was followed by the development of the laser. This made available for the first time a powerful source of highly coherent light and made it possible to record holograms of diffusely reflecting objects with appreciable depth [Leith & Upatnieks, 1964].

Almost at the same time, another major advance in holography was reported by Denisyuk [1962, 1963, 1965]. In his technique, which has some similarities to Lippmann's technique of colour photography, the object and reference waves are incident on the photographic emulsion from opposite sides. As a result, the interference fringes recorded are actually layers, almost parallel to the surface of the emulsion and about half a wavelength apart. Such holograms, when illuminated with white light from a point source, selectively reflect only a narrow wavelength band to reconstruct a monochromatic image.

1.3.   The development of optical holography

These advances set off an explosive growth of activity and optical holography soon found a very large number of scientific applications. These included high-resolution imaging of aerosols [Thompson, Ward & Zinky, 1967], imaging through diffusing and aberrating media [Kogelnik, 1965; Leith & Upatnieks, 1966], multiple imaging [Lu, 1968; Groh, 1968], computer generated holograms [Lohmann & Paris, 1967] and the production and correction of optical elements [Upatnieks, Vander Lugt & Leith, 1966]. Other applications were related to information storage and information processing and included image deblurring [Stroke, Restrick, Funkhouser & Brumm, 1965] and character recognition [Vander Lugt, Rotz & Klooster, 1965].

Perhaps the most significant of these applications was holographic interferometry, which was discovered almost simultaneously by several groups, all working independently [Brooks, Heflinger & Wuerker, 1965; Burch, 1965; Collier, Doherty & Pennington, 1965; Haines and Hildebrand, 1965; Powell & Stetson, 1965]. The technical advance it represented was astonishing. It became possible, for the first time, as shown in fig. 1.5, to map the displacements of a relatively rough surface with an accuracy of a

Fig. 1.5. Holographic interferogram of a XVth century painting, revealing separation of the surface layers [Amadesi, Gori, Grella & Guattari, 1974].

fraction of a micrometre; it was even possible to make interferometric comparisons of stored wavefronts that existed at different times.

In the field of three-dimensional displays, quite spectacular developments followed, including life-size holograms, and portraits with pulsed lasers (see figs. 1.6 and 1.7), as well as multicolour images. However, these represented a scientific *tour-de-force* rather than a viable technology, and little further progress was made until Benton [1969] invented the rainbow

Fig. 1.6. The Venus de Milo; this hologram (1.5 m × 1.0 m) was produced by J. M. Fournier and G. Tribillon at the Laboratoire de Physique Général et Optique, Université de Besançon, France, in 1976.

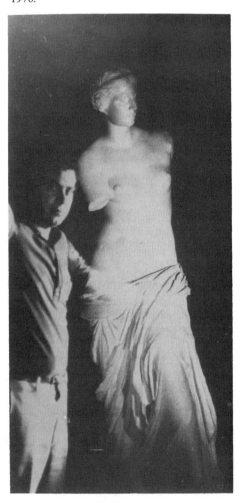

hologram. This was a transmission hologram in which vertical parallax was sacrificed to gain two major advantages – it could be illuminated with white light and it gave a very bright monochromatic image.

While it took some time for the practical advantages of the rainbow hologram to be appreciated, it ultimately resulted in developments in two areas. One was a series of new techniques for multicolour and achromatic imaging, while the other was the white-light holographic stereogram of Cross [see Benton, 1975]. The latter is based on earlier work in which a three-dimensional image was built up from holograms of a number of views of an object from different angles in the horizontal plane, but can be viewed in white light, since it makes use of rainbow holograms. Because they are well adapted to commercial displays, white-light holographic stereograms now command a substantial market.

Another significant area of progress has been that of recording materials. Fine-grain silver halide photographic emulsions are still the most widely used recording medium, and considerable work has been done to improve their characteristics as well as to optimize processing techniques. In addition, other materials such as dichromated gelatin, photothermoplastics

Fig. 1.7. Professor Gabor with his holographic portrait; this hologram was produced by R. Rinehart at the McDonnel Douglas Electronics Company in 1971, using a pulsed laser.

and photorefractive crystals are now being used, to an increasing extent, for specific applications for which they offer definite advantages. Very interesting possibilities have been opened up in dynamic holography, including the generation of phase-conjugate waves by Brillouin scattering, and holography in resonant media [Denisyuk, 1980].

The striking realism of holographic images has always been something that has fascinated scientists as well as laymen. However, as a result of the developments of the last decade, holography has ceased to be a novelty and has become a well-established technique with many invaluable applications. As always in scientific research, some of these advances were essentially unanticipated; we can fully expect many more such surprises in the next few years.

2

The wavefront reconstruction process

The concepts of holography outlined in Chapter 1 can now be formulated and discussed in more specific terms.

2.1. The in-line (Gabor) hologram

Consider the optical system shown in fig. 2.1, which is essentially that used by Gabor [1948] to demonstrate holographic imaging. In this setup, the coherent light source as well as the object, which is a transparency containing small opaque details on a clear background, are located along the axis normal to the photographic plate.

When the object is illuminated with a uniform parallel beam, the light transmitted by it consists of two parts. The first is a relatively strong, uniform plane wave corresponding to the directly transmitted light. This constitutes the reference wave, and, since its amplitude and phase do not vary across the photographic plate, its complex amplitude can be written as a real constant r. The second is a weak scattered wave due to the transmittance variations in the object. The complex amplitude of this wave, which varies across the photographic plate, can be written as $o(x, y)$, where $|o(x, y)| \ll r$.

The resultant complex amplitude at any point on the photographic plate

Fig. 2.1. Optical system for recording an in-line (Gabor) hologram.

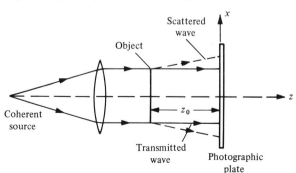

is the sum of these two, so that the intensity at this point (see Appendix A3.1) is

$$I(x, y) = |r + o(x, y)|^2,$$
$$= r^2 + |o(x, y)|^2 + r\, o(x, y) + r\, o^*(x, y), \qquad (2.1)$$

where $o^*(x, y)$ is the complex conjugate of $o(x, y)$.

A positive transparency is made from this recording. For simplicity, it is assumed that this has been processed so that its amplitude transmittance **t** (the ratio of the transmitted amplitude to that incident on it) is a linear function of the intensity and can be written as

$$\mathbf{t} = \mathbf{t}_0 + \beta TI, \qquad (2.2)$$

where $\mathbf{t}_0$ is a constant background transmittance, T is the exposure time and β is a parameter determined by the photographic material used and the processing conditions. The amplitude transmittance of the transparency is, accordingly,

$$\mathbf{t}(x, y) = \mathbf{t}_0 + \beta T[r^2 + |o(x, y)|^2 + r\, o(x, y)$$
$$+ r\, o^*(x, y)]. \qquad (2.3)$$

To view the reconstructed image, this transparency is replaced in the same position as the original photographic plate and illuminated once again with the same parallel beam of monochromatic light used to record the hologram, as shown in fig. 2.2. Since the complex amplitude at any point in this beam is, apart from a constant factor, the same as that in the reference beam, the complex amplitude transmitted by the hologram can be written as

$$u(x, y) = r\, \mathbf{t}(x, y),$$
$$= r(\mathbf{t}_0 + \beta Tr^2) + \beta Tr|o(x, y)|^2$$
$$+ \beta Tr^2 o(x, y) + \beta Tr^2 o^*(x, y). \qquad (2.4)$$

Fig. 2.2. Optical system for reconstructing the image from an in-line (Gabor) hologram, showing the formation of the twin images.

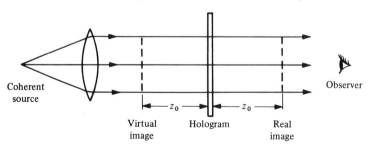

Coherent source

Observer

z_0 z_0

Virtual image   Hologram   Real image

The complex amplitude of the transmitted wave consists, therefore, of four distinct components.

The first of these, $r(t_0 + \beta T r^2)$, is a uniformly attenuated plane wave, the directly transmitted beam.

The second term, $\beta T r |o(x, y)|^2$, is extremely small in comparison to the other terms, since it has been assumed initially that $|o(x, y)| \ll r$. Under these conditions, it can be neglected.

The third term, $\beta T r^2 o(x, y)$, is, except for a constant factor, the same as the original wavefront from the object incident on the photographic plate. Hence, it gives rise to a reconstructed image of the object in its original position. Since this is located behind the transparency at a distance z_0 from it, and the reconstructed wave appears to diverge from it, it is a virtual image.

In the same manner, the fourth term corresponds to a wavefront which resembles the wavefront diverging from the object, except that it is of opposite curvature. Hence, it converges to form a real image, the conjugate image, at the same distance z_0 in front of the hologram.

It is apparent that, with such a system, an observer focusing on one image sees it superposed on the out-of-focus twin image as well as a strong coherent background. This constitutes its most serious limitation.

Another major limitation is the need for the object to have a high average transmittance, if the second term on the right hand side of (2.4), which has been taken as negligible, is not to interfere with the reconstructed image. Typically, it is possible to form images of fine opaque lines on a transparent background, but not *vice versa*.

Finally, it should be noted that the hologram used to reconstruct the image must be a positive transparency. Since the image-forming waves interfere with the background in the process of reconstruction, if the plate which has been exposed in the recording step is used directly (in which case β in (2.2) is negative), the reconstructed image will also be a negative.

2.2. The off-axis (Leith–Upatnieks) hologram

The first successful technique for separating the twin images was developed by Leith & Upatnieks [1962, 1963, 1964]. As shown in fig. 2.3, a separate reference beam derived from the same coherent source is allowed to fall on the photographic plate, during the recording process, at an offset angle θ to the beam from the object. For simplicity, this reference beam can be assumed to be a collimated beam of uniform intensity.

The complex amplitude due to the object beam at any point (x, y) on the photographic plate can then be written as

$$o(x, y) = |o(x, y)| \exp[-i\phi(x, y)], \tag{2.5}$$

while that due to the reference beam is

$$r(x, y) = r \exp(i2\pi\xi_r x), \tag{2.6}$$

where $\xi_r = (\sin\theta)/\lambda$, since the reference beam has uniform intensity and only its phase varies across the photographic plate.

The resultant intensity at the photographic plate is

$$
\begin{aligned}
I(x, y) &= |r(x, y) + o(x, y)|^2 \\
&= |r(x, y)|^2 + |o(x, y)|^2 \\
&\quad + r|o(x, y)| \exp[-i\phi(x, y)] \exp(-i2\pi\xi_r x) \\
&\quad + r|o(x, y)| \exp[i\phi(x, y)] \exp(i2\pi\xi_r x), \\
&= r^2 + |o(x, y)|^2 + 2r|o(x, y)| \cos[2\pi\xi_r x + \phi(x, y)]. \tag{2.7}
\end{aligned}
$$

As can be seen from (2.7), the amplitude and phase of the object wave are encoded, respectively, as amplitude and phase modulation of a set of interference fringes equivalent to a spatial carrier with a spatial frequency equal to ξ_r.

If, as in (2.2), it is assumed that the amplitude transmittance of the photographic plate after processing is linearly related to the intensity in the interference pattern, the amplitude transmittance of the hologram can be written as

$$
\begin{aligned}
t(x, y) = t_0 + \beta T\{ &|o(x, y)|^2 \\
&+ r|o(x, y)| \exp[-i\phi(x, y)] \exp(-i2\pi\xi_r x) \\
&+ r|o(x, y)| \exp[i\phi(x, y)] \exp(i2\pi\xi_r x)]\}, \tag{2.8}
\end{aligned}
$$

where β is the slope (in this case, negative) of the amplitude transmittance *versus* exposure characteristic of the photographic material, T is the exposure time and t_0 is a constant background transmittance.

To reconstruct the image, the hologram is illuminated once again, as shown in fig. 2.4 with the same reference beam as was used to record it. The

Fig. 2.3. Hologram recording with an off-axis reference beam.

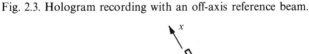
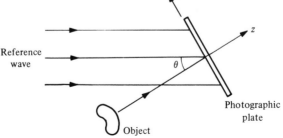

complex amplitude $u(x, y)$ of the transmitted wave is, in this case also, the sum of four terms, each corresponding to one of the terms of (2.8), and can be written as

$$u(x, y) = r(x, y)t(x, y),$$
$$= u_1(x, y) + u_2(x, y) + u_3(x, y) + u_4(x, y), \qquad (2.9)$$

where

$$u_1(x, y) = t_0 r \exp(i2\pi\xi_r x), \qquad (2.10)$$

$$u_2(x, y) = \beta T r |o(x, y)|^2 \exp(i2\pi\xi_r x), \qquad (2.11)$$

$$u_3(x, y) = \beta T r^2 o(x, y), \qquad (2.12)$$

$$u_4(x, y) = \beta T r^2 o^*(x, y) \exp(i4\pi\xi_r x). \qquad (2.13)$$

The first term on the right hand side of (2.9), $u_1(x, y)$, is, as before, merely the attenuated reference beam, which is a plane wave directly transmitted through the hologram. This directly transmitted beam is surrounded by a halo due to the second term, $u_2(x, y)$, which is spatially varying. The angular spread of this halo is determined by the angular extent of the object.

The third term, $u_3(x, y)$ is identical with the original object wave, except for a constant factor, and generates a virtual image of the object in its original position; this wave makes an angle θ with the directly transmitted wave. Similarly, the fourth term, $u_4(x, y)$, gives rise to the conjugate real image. However, in this case, the fourth term includes an exponential factor, $\exp(i4\pi\xi_r x)$, which indicates that the conjugate wave is deflected off the axis at an angle approximately twice that which the reference wave makes with it.

Thus, even though two images – one real and one virtual – are still reconstructed in this setup, they are angularly separated from the directly

Fig. 2.4. Image reconstruction by a hologram recorded with an off-axis reference beam.

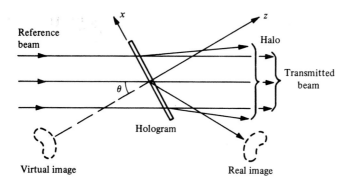

transmitted beam and from each other, and if the offset angle θ of the reference beam is made large enough, it is possible to ensure that there is no overlap. This method therefore eliminates at one stroke all the major drawbacks of Gabor's original in-line arrangement. In addition, it is also interesting that the sign of β in this case only affects the phase of the reconstructed image – a 'positive' image is obtained even if the hologram recording is a photographic negative.

The minimum value of the offset angle θ required to ensure that each of the images can be observed without any interference from its twin, as well as from the directly transmitted beam and the halo of scattered light surrounding it, is determined by the minimum spatial carrier frequency (ξ_r) for which there is no overlap between the angular spectra of the third and fourth terms and those of the first and second terms.

These angular spectra are the Fourier transforms (see Appendix A1) of these terms and can be written as follows.

$$U_1(\xi, \eta) = \mathscr{F}\{t_0 r \exp{(i2\pi\xi_r x)}\},$$
$$= t_0 r \delta(\xi + \xi_r, \eta). \tag{2.14}$$

$$U_2(\xi, \eta) = \mathscr{F}\{\beta Tr|o(x, y)|^2 \exp{(i2\pi\xi_r x)}\},$$
$$= \beta Tr[O(\xi, \eta) \star O(\xi, \eta) * \delta(\xi + \xi_r, \eta)], \tag{2.15}$$

where $O(\xi, \eta) = \mathscr{F}\{o(x, y)\}$ is the spatial frequency spectrum of the object beam, and the symbols $\star$ and $*$ denote, respectively, the operations of correlation and convolution.

$$U_3(\xi, \eta) = \mathscr{F}\{\beta Tr^2 o(x, y)\},$$
$$= \beta Tr^2 O(\xi, \eta). \tag{2.16}$$

$$U_4(\xi, \eta) = \mathscr{F}\{\beta Tr^2 o^*(x, y) \exp{(i4\pi\xi_r x)}\},$$
$$= \beta Tr^2 O^*(\xi, \eta) * \delta(\xi + 2\xi_r, \eta). \tag{2.17}$$

As can be seen from fig. 2.5, which shows these spectra schematically, the term $|U_3|$ is merely the object-beam spectrum multiplied by a constant and is centred at the origin of the spatial frequency plane. The term $|U_1|$ corresponds to the spatial frequency of the carrier fringes and is a δ function located at $(-\xi_r, 0)$, while $|U_2|$ is centred on this δ function and, being proportional to the auto-correlation function of $O(\xi, \eta)$, has twice the extent of the object-beam spectrum. Finally, $|U_4|$ is similar to $|U_3|$ but is displaced to a centre frequency $(-2\xi_r, 0)$.

Evidently, $|U_3|$ and $|U_4|$ will not overlap $|U_1|$ and $|U_2|$ if the spatial carrier frequency ξ is chosen so that

$$\xi_r \geqslant 3\xi_m, \tag{2.18}$$

where ξ_m is the highest frequency in the spatial frequency spectrum of the object beam.

2.3.   Fourier holograms

Another interesting category of hologram recording configurations is one in which the waves that interfere at the hologram plane are the Fourier transforms of the object and reference waves. Normally, this implies an object which lies in a single plane, or is of limited thickness.

A typical optical arrangement for recording such a hologram (Vander Lugt, 1964) is shown in fig. 2.6. The object, a transparency located in the front focal plane of a lens, is illuminated by a parallel beam of monochromatic light. If the complex amplitude leaving the object plane is $o(x, y)$, the complex amplitude of the object wave at the photographic plate located in the back focal plane of the lens is

$$O(\xi, \eta) = \mathscr{F}\{o(x, y)\}. \qquad (2.19)$$

The reference beam is derived from a point source also located in the front focal plane of the lens. If $\delta(x + b, y)$ is the complex amplitude of this point

Fig. 2.5. Spatial frequency spectra of (*a*) the object beam and (*b*) a hologram recorded with an off-axis reference beam.

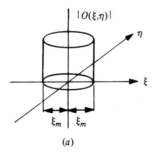

(*a*)

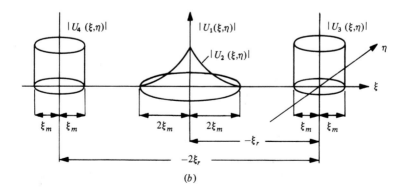

(*b*)

source, the complex amplitude of the reference wave at the hologram plane is

$$R(\xi,\eta) = \exp(-i2\pi\xi b). \tag{2.20}$$

The intensity in the interference pattern formed by these two waves is therefore

$$I(\xi, \eta) = 1 + |O(\xi, \eta)|^2 + O(\xi, \eta)\exp(i2\pi\xi b)$$
$$+ O^*(\xi, \eta)\exp(-i2\pi\xi b). \tag{2.21}$$

To reconstruct the image, the processed hologram is placed in the front focal plane of the lens and illuminated with a parallel beam of monochromatic light of unit amplitude as shown in fig. 2.7. If it is assumed, as before, that the amplitude transmittance of the processed hologram is a linear function of $I(\xi, \eta)$, the complex amplitude of the light transmitted by the hologram is

$$U(\xi, \eta) = \mathbf{t}_0 + \beta T I(\xi, \eta). \tag{2.22}$$

The complex amplitude in the back focal plane of the lens is then the

Fig. 2.6. Optical system for recording a Fourier hologram.

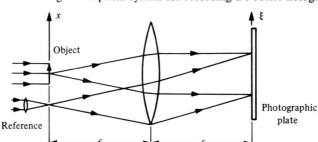

Fig. 2.7. Reconstruction setup for a Fourier hologram.

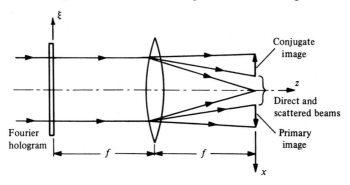

Fourier transform of $U(\xi, \eta)$,

$$u(x, y) = \mathscr{F}\{U(\xi, \eta)\},$$
$$= (\mathbf{t}_0 + \beta T)\delta(x, y) + \beta To(x, y) \star o(x, y)$$
$$+ \beta To(x - b, y) + \beta To^*(-x + b, -y). \tag{2.23}$$

As shown in fig. 2.7, the first term on the right-hand side of (2.23) comes to a focus on the axis, while the second term forms a halo around it. The third term corresponds to the original object wave shifted downwards by a distance b, while the fourth term is the conjugate of the original object wave inverted and shifted upwards by the same amount b. Both the images are real and can be recorded on a photographic film placed in the back focal plane of the lens. Since the film records only the intensity distribution in the image, it is possible, in this case, to identify the conjugate image only by the fact that it is inverted (see fig. 2.8).

Fourier holograms have the useful property that the reconstructed image is stationary even when the hologram is translated in its own plane. This is because a shift of a function in the spatial domain only results in its Fourier transform being multiplied by a phase factor which is a linear function of the spatial frequency (see Appendix A1). This phase factor has no effect on the intensity distribution in the image.

2.4 The lensless Fourier hologram

A hologram with the same properties as a Fourier hologram can be obtained even without a lens to produce the Fourier transform of the object wave, provided the reference wave is produced by a point source in the plane of the object [Stroke, 1965; Stroke, Brumm & Funkhouser, 1965].

Fig. 2.8. Twin images reconstructed by a Fourier hologram.

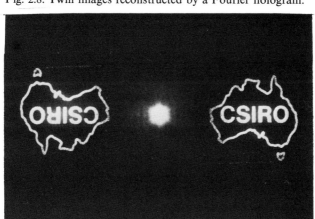

Consider the setup shown in fig. 2.9 in which, again, the object is a transparency illuminated with a plane wave, and let the complex amplitude of the wave leaving the object plane be $o(x_1, y_1)$. It can then be shown using the Fresnel–Kirchhoff integral (see Appendix A2) that the complex amplitude at the photographic plate due to this wave can be written as

$$o(x_2, y_2) = (i/\lambda z_0) \exp\left[-(i\pi/\lambda z_0)(x_2^2 + y_2^2) \right] O(\xi, \eta). \tag{2.24}$$

where z_0 is the distance from the object plane to the hologram plane, $\xi = x_2/\lambda z_0$, $\eta = y_2/\lambda z_0$, and

$$O(\xi, \eta) = \mathscr{F}\{ o(x_1, y_1) \exp\left[-(i\pi/\lambda z_0)(x_1^2 + y_1^2) \right] \}. \tag{2.25}$$

This is essentially the Fourier transform of the object wave, modified by a spherical phase factor which depends on the distance from the object to the hologram. Similarly, it can be shown that the complex amplitude at the photographic plate due to the reference wave is

$$r(x_2, y_2) = r \exp\left[-(i\pi/\lambda z_0)(x_2^2 + y_2^2) \right] \exp(-i2\pi\xi b), \tag{2.26}$$

where b is the distance of the reference point source from the z axis.

The intensity in the resulting interference pattern is then

$$\begin{aligned}
I(x_2, y_2) = &\, r^2 + |o(x_2, y_2)|^2 \\
&+ (i/\lambda z_0) O(\xi, \eta) \exp(i2\pi\xi b) \\
&+ (i/\lambda z_0) O^*(\xi, \eta) \exp(-i2\pi\xi b).
\end{aligned} \tag{2.27}$$

A comparison with (2.21) shows that this is very similar to that obtained with the conventional setup used to produce a Fourier hologram, and the resulting hologram has essentially the same properties.

It is apparent that, in this recording configuration, the effect of the spherical phase factor associated with the near-field or Fresnel diffraction

Fig. 2.9. Optical system for recording a lensless Fourier hologram.

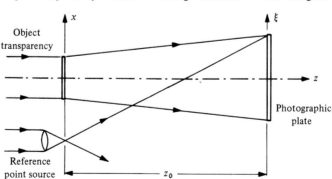

pattern of the object transparency is eliminated by the use of a spherical reference wave with the same curvature.

2.5.   Image holograms

Significant advantages result when a hologram is recorded of a real image of the object instead of the object itself [Rosen, 1966; Stroke, 1966]. A typical optical arrangement, in which a large lens is used to project a real image of the object is shown in fig. 2.10.

With such a setup it is possible to position the projected image of the object, when recording the hologram, on the photographic plate (straddling it). The reconstructed primary image is then formed in the same position, so that one half of it appears to be behind the plate while the other half is in front of it.

As will be shown later (see section 3.5.3), such an image hologram has the very useful property that it can be illuminated with an incoherent source of appreciable size and spectral bandwidth and will still produce an acceptably sharp image. Another advantage of such a hologram (see section 3.6.2) is increased image luminance; however, this is offset partially by the fact that the angle over which the image can be viewed is limited by the angular aperture of the imaging lens.

2.6.   Fraunhofer holograms

One of the first applications of optical holography arose from the recognition of a situation in which in-line holography could be used

Fig. 2.10. Formation of an image hologram.

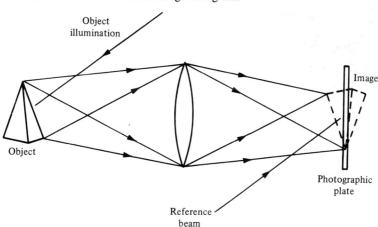

Fig. 2.11. Formation of a Fraunhofer hologram.

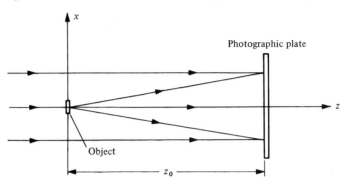

without the usual problems associated with the presence of the conjugate image. This led to the development of the Fraunhofer hologram.

Such a hologram is formed when, as shown in fig. 2.11, the object is small enough for its far-field diffraction pattern to be formed at the photographic plate. For this, z_0, the distance of the plate from the object, must satisfy the far-field condition (see Appendix A2)

$$z_0 \gg (x_0^2 + y_0^2)/\lambda,$$

where x_0 and y_0 define the maximum dimensions of the object.

A typical case is a hologram of a group of point scatterers such as aerosol particles [Thompson, 1963]. The particles are so small that even at distances of a few millimetres the hologram can be considered to lie in the far field of any individual particle. In such a case, the diffracted light due to the conjugate image is spread over such a large area in the primary image plane that it only contributes a weak, uniform background. As a result the image can be viewed without interference from its conjugate.

3

The reconstructed image

This chapter discusses the characteristics of the reconstructed image and the dependence of these characteristics on various parameters of the optical system used for recording and reconstruction.

3.1. Images of a point

To simplify the analysis it is convenient to consider the hologram of a point object O whose coordinates are (x_O, y_O, z_O) recorded with a reference wave from a point source R located at (x_R, y_R, z_R), as shown in fig. 3.1(a) [Meier, 1965].

The complex amplitude of the object wave at a point H in the hologram plane with coordinates (x_H, y_H, z_H) can be written as $a_O = |a_O| \exp(-i\phi_O)$, where ϕ_O is the phase of the wave at this point relative to that at O. The phase ϕ_O can be computed from the optical path difference (see Appendix A2) and is given, to a first-order approximation, by the expression

$$\phi_O = (\pi/\lambda_1)[(1/z_O)(x_H^2 + y_H^2 - 2x_H x_O - 2y_H y_O)], \tag{3.1}$$

where λ_1 is the wavelength of the light used to record the hologram.

Similarly, the complex amplitude of the reference wave at the point H (x_H, y_H, z_H) can be written as $a_R = |a_R| \exp(-i\phi_R)$, where

$$\phi_R = (\pi/\lambda_1)[(1/z_R)(x_H^2 + y_H^2 - 2x_H x_R - 2y_H y_R)]. \tag{3.2}$$

As in the case of (2.7), the position and spacing of the interference fringes in the hologram produced by these two waves are determined by the phase difference $(\phi_R - \phi_O)$.

We shall assume that the processed hologram is illuminated by monochromatic light of wavelength λ_2 from a point source P located at (x_P, y_P, z_P), as shown in fig. 3.1(b). In the same manner, the complex amplitude of this wave at the point H (x_H, y_H, z_H) in the hologram plane can be written as $a_P = |a_P| \exp|(-i\phi_P)$, where

$$\phi_P = (\pi/\lambda_2)[(1/z_P)(x_H^2 + y_H^2 - 2x_H x_P - 2y_H y_P)]. \tag{3.3}$$

If we assume linear recording (see 2.2 and 2.3), the terms of the

23

transmitted wavefront giving rise to the two reconstructed images (see 2.12–2.13) are then, apart from a constant factor,

$$u_3 = a_P a_R^* a_O,$$
$$= |a_P| |a_R| |a_O| \exp[-i(\phi_P - \phi_R + \phi_O)], \tag{3.4}$$

and

$$u_4 = a_P a_R a_O^*,$$
$$= |a_P| |a_R| |a_O| \exp[-i(\phi_P + \phi_R - \phi_O)]. \tag{3.5}$$

If u_3 is written in the form $u_3 = |u_3| \exp(-i\phi_3)$, it follows that the phase ϕ_3 of this reconstructed wave at H (x_H, y_H, z_H) is

$$\phi_3 = \phi_P - \phi_R + \phi_O. \tag{3.6}$$

When we substitute the values of ϕ_O, ϕ_R and ϕ_P from (3.1), (3.2) and (3.3) in (3.6), and set $(\lambda_2/\lambda_1) = \mu$, we have

$$\phi_3 = (\pi/\lambda_2)\left[(x_H^2 + y_H^2)\left(\frac{1}{z_P} + \frac{\mu}{z_O} - \frac{\mu}{z_R}\right) \right.$$
$$\left. - 2x_H\left(\frac{x_P}{z_P} + \frac{\mu x_O}{z_O} - \frac{\mu x_R}{z_R}\right) - 2y_H\left(\frac{y_P}{z_P} + \frac{\mu y_O}{z_O} - \frac{\mu y_R}{z_R}\right) \right]. \tag{3.7}$$

Fig. 3.1. Coordinate system used to study image formation by a hologram.

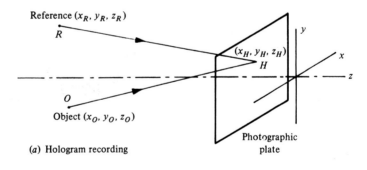

Reference (x_R, y_R, z_R)

R

(x_H, y_H, z_H)

H

O

Object (x_O, y_O, z_O)

Photographic plate

(*a*) Hologram recording

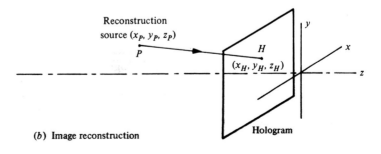

Reconstruction source (x_P, y_P, z_P)

P

H

(x_H, y_H, z_H)

(*b*) Image reconstruction

Hologram

Since we require u_3 to produce a point image, u_3 must be a spherical wave, in which case it should be possible to express its phase ϕ_3 at H (x_H, y_H, z_H) in the form

$$\phi_3 = (\pi/\lambda_2)[(1/z_3)(x_H^2 + y_H^2 - 2x_H x_3 - 2y_H y_3)], \tag{3.8}$$

where (x_3, y_3, z_3) are the coordinates of this point image.

Hence, if we equate the coefficients of similar terms in (3.7) and (3.8), the coordinates of the image formed by u_3 can be written as

$$x_3 = \frac{x_P z_O z_R + \mu x_O z_P z_R - \mu x_R z_P z_O}{z_O z_R + \mu z_P z_R - \mu z_P z_O}, \tag{3.9}$$

$$y_3 = \frac{y_P z_O z_R + \mu y_O z_P z_R - \mu y_R z_P z_O}{z_O z_R + \mu z_P z_R - \mu z_P z_O}, \tag{3.10}$$

$$z_3 = \frac{z_P z_O z_R}{z_O z_R + \mu z_P z_R - \mu z_P z_O}. \tag{3.11}$$

Similarly, the coordinates of the image formed by the conjugate wave u_4 can be written as

$$x_4 = \frac{x_P z_O z_R - \mu x_O z_P z_R + \mu x_R z_P z_O}{z_O z_R - \mu z_P z_R + \mu z_P z_O}, \tag{3.12}$$

$$y_4 = \frac{y_P z_O z_R - \mu y_O z_P z_R + \mu y_R z_P z_O}{z_O z_R - \mu z_P z_R + \mu z_P z_O}, \tag{3.13}$$

$$z_4 = \frac{z_P z_O z_R}{z_O z_R - \mu z_P z_R + \mu z_P z_O}. \tag{3.14}$$

3.2. Image magnification

We can consider an extended object to be made up of a number of point objects, and apply the preceding analysis to evaluate the form and other characteristics of the reconstructed image.

3.2.1. Lateral magnification

The lateral magnification of the primary image can be defined either as

$$M_{\text{lat}, 3} = (dx_3/dx_O), \tag{3.15}$$

or as

$$M_{\text{lat}, 3} = (dy_3/dy_O). \tag{3.16}$$

Both these definitions lead to the same result, since from (3.9) and (3.10),

$$M_{\text{lat}, 3} = \left[1 + z_O\left(\frac{1}{\mu z_P} - \frac{1}{z_R}\right)\right]^{-1}. \tag{3.17}$$

Similarly, for the conjugate image, we have

$$M_{\text{lat}, 4} = (dx_4/dx_O), \tag{3.18}$$

or

$$M_{\text{lat}, 4} = (dy_4/dy_O), \tag{3.19}$$

which, from (3.12) and (3.13), yield

$$M_{\text{lat},4} = \left[1 - z_O\left(\frac{1}{\mu z_P} + \frac{1}{z_R}\right)\right]^{-1}. \tag{3.20}$$

3.2.2. Angular magnification

If we assume that the observer's eye is located in the hologram plane, the angular magnification of the primary image can be defined as

$$M_{\text{ang}} = \frac{d(x_3/z_3)}{d(x_O/z_O)}, \tag{3.21}$$

which reduces to

$$|M_{\text{ang}}| = \mu, \tag{3.22}$$

and is the same as that for the conjugate image.

3.2.3. Longitudinal magnification

The longitudinal magnification of the primary image can be calculated from the relation

$$
\begin{aligned}
M_{\text{long}, 3} &= (dz_3/dz_O), \\
&= \frac{1}{\mu}\frac{d}{dz_O}\left\{\frac{z_O}{1 + z_O[(1/\mu z_P) - (1/z_R)]}\right\}, \\
&= \frac{1}{\mu}\left\{\frac{1}{1 + z_O[(1/\mu z_P) - (1/z_R)]}\right\}^2, \\
&= \frac{1}{\mu}M_{\text{lat}, 3}^2.
\end{aligned} \tag{3.23}
$$

Similarly, the longitudinal magnification of the conjugate image is

$$
\begin{aligned}
M_{\text{long}, 4} &= (dz_4/dz_O), \\
&= -\frac{1}{\mu}\frac{d}{dz_O}\left\{\frac{z_O}{1 - z_O[(1/\mu z_P) + (1/z_R)]}\right\}, \\
&= -\frac{1}{\mu}\left\{\frac{1}{1 - z_O\{(1/\mu z_P) + (1/z_R)]}\right\}^2, \\
&= -\frac{1}{\mu}M_{\text{lat}, 4}^2.
\end{aligned} \tag{3.24}
$$

Note that $M_{\text{long},3}$ and $M_{\text{long},4}$ have opposite signs. The consequences of this are discussed in the next section.

3.3. Orthoscopic and pseudoscopic images

To understand the implications of the opposite signs of $M_{\text{long},3}$ and $M_{\text{long},4}$, consider an off-axis hologram recorded with a collimated reference beam incident normally to the photographic plate, as shown in fig. 3.2(a). When this hologram is illuminated once again with the same collimated reference beam, as shown in fig. 3.2(b), it reconstructs two images, one

Fig. 3.2. Formation of orthoscopic and pseudoscopic images by a hologram.

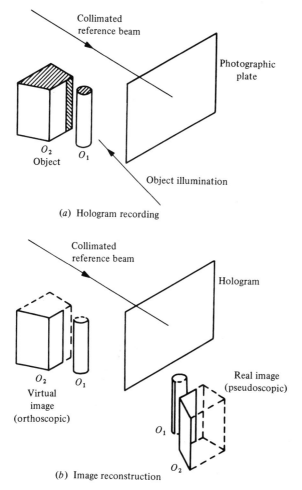

Collimated reference beam

Photographic plate

O_2
Object

O_1

Object illumination

(*a*) Hologram recording

Collimated reference beam

Hologram

O_2
Virtual image
(orthoscopic)

O_1

Real image
(pseudoscopic)

O_1

O_2

(*b*) Image reconstruction

virtual and the other real, both of which, at first sight, are exact replicas of the object. However, the two images differ in one very important respect.

Thus, while the virtual image is located in the same position as the object and exhibits the same parallax properties, the real image is formed at the same distance from the hologram but in front of it. Since, from (3.11) and (3.14), corresponding points on the real and virtual images are located at equal distances from the plane of the hologram, the real image has the curious property that its depth is inverted. Such an image is not formed with a normal optical system; it is therefore called a pseudoscopic image as opposed to a normal or orthoscopic image, [Leith & Upatnieks, 1964; Rosen, 1967].

This depth inversion results in conflicting visual clues which make viewing of the real image psychologically unsatisfactory. Thus, if O_1 and O_2 are two elements in the object field, and if O_1 blocks the light scattered by O_2 at a certain angle, the hologram records information only on the element O_1 at this angle and records no information about this part of O_2. An observer viewing the real image from the corresponding direction then cannot see this part of O_2, which, contrary to normal experience, is obscured by O_1, even though O_2 is in front of O_1.

3.3.1.   Production of an orthoscopic real image

An orthoscopic real image of an object can be produced by recording two holograms in succession [Rotz & Friesem, 1966].

In the first step, as shown in fig. 3.3, a hologram is recorded of the object with a collimated reference beam. When this hologram is illuminated once again with the collimated reference beam used to record it, it reconstructs two images of the object at unit magnification, one of them being an orthoscopic virtual image, while the other is a pseudoscopic real image. A second hologram is then recorded of this real image with a second collimated reference beam.

When the second hologram is illuminated with a collimated beam it reconstructs a pseudoscopic virtual image located in the same position as the real image formed by the primary hologram. However, the real image formed by the second hologram is an orthoscopic image. Since collimated reference beams are used throughout, the final real image is the same size as the original object and free from aberrations.

A simpler alternative is to record a hologram of an orthoscopic real image of the object formed by a lens, as shown in fig. 3.4. Alternatively, a concave mirror can be used. When this hologram is illuminated with the original reference wave, it reconstructs the original object wave which gives rise to an orthoscopic real image.

3.4.   Image aberrations

If the processed hologram is replaced in exactly the same position in the setup in which it was recorded and illuminated once again with the

Fig. 3.3. Production of an orthoscopic real image by recording two holograms in succession [Rotz & Friesem, 1966].

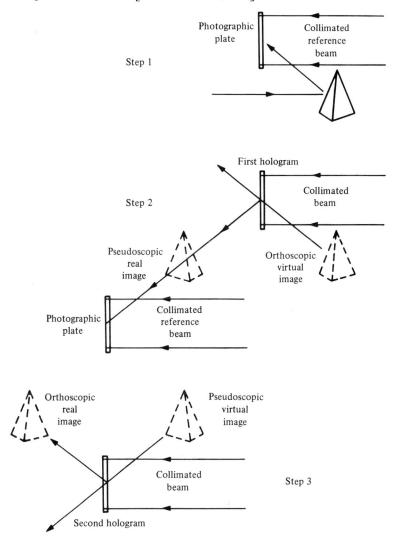

reference beam used to record it, ($z_P = z_R$, $\mu = 1$), the primary image coincides exactly with the object. In any other case, the image may exhibit aberrations.

Thus, while the locations of the image points were calculated in section 3.1 on the assumption that the reconstructed wavefronts were spherical, this assumption is not strictly true. The aberrations of these wavefronts can then be defined as the phase differences between the reference spheres centred on these points and the actual wavefronts in the hologram plane.

Hologram aberrations can be classified exactly like lens aberrations [Hopkins, 1950]. It is convenient to use polar coordinates (ρ, θ) in the hologram plane, instead of the cartesian coordinates (x_H, y_H). The third-order aberration can then be written as

$$
\begin{aligned}
\Delta\phi_3 = (2\pi/\lambda_2)[&-(1/8)\rho^4 S \\
&+(1/2)\rho^3(C_x \cos\theta + C_y \sin\theta) \\
&-(1/2)\rho^2(A_x \cos^2\theta + A_y \sin^2\theta \\
&\qquad\qquad +2A_x A_y \cos\theta \sin\theta) \\
&-(1/4)\rho^2 F \\
&+(1/2)\rho(D_x \cos\theta + D_y \sin\theta)],
\end{aligned}
\tag{3.25}
$$

where S is the coefficient of spherical aberration, C_x and C_y are the coma coefficients, A_x and A_y are the coefficients of astigmatism, F is the coefficient for the field curvature and D_x and D_y are the distortion coefficients.

If we retain the third degree terms in the expansion for the phase of a spherical wavefront, it is possible to calculate these coefficients [Meier, 1965; Leith, Upatnieks & Haines, 1965]. In the discussion that follows, only

Fig. 3.4. Production of a hologram which gives an orthoscopic real image using a lens.

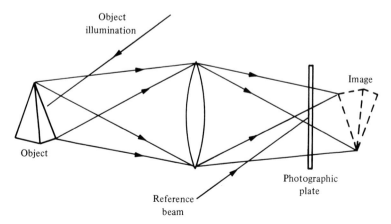

the conjugate (real) image will be considered, and it will be assumed for simplicity that the object lies on the x axis ($y_O = 0$). The expressions for the aberration coefficients of the primary (virtual) image can be obtained by changing the signs of z_O and z_R in the corresponding expressions for the conjugate image.

Enlarging or reducing the hologram can make it easier to eliminate some of the aberrations. However, this possibility is not taken into account here, since it is impracticable for a normal off-axis hologram because of the very close spacing of the carrier fringes.

3.4.1. Spherical aberration

It can be shown that the coefficient of spherical aberration is

$$S = (1/z_P^3) - (\mu/z_O^3) + (\mu/z_R^3) - (1/z_4^3). \tag{3.26}$$

When $z_R = z_O$, $z_3 = z_P$, and the spherical aberration disappears for both the reconstructed wavefronts. The reason for this is easily understood: because the two interfering wavefronts have the same curvature, the phase difference $(\phi_O - \phi_R)$ between them changes linearly across the field. The resulting straight hologram fringes cannot introduce spherical aberration.

3.4.2. Coma

The coefficient of coma is

$$C_x = (x_P/z_P^3) - (\mu x_O/z_O^3) + (\mu x_R/z_R^3) - (x_4/z_4^3). \tag{3.27}$$

Coma can be eliminated only if $z_R = z_O$ and $z_P = \pm z_O$: it then disappears for both images.

3.4.3. Astigmatism

The coefficient of astigmatism is

$$A_x = (x_P^2/z_P^3) - (\mu x_O^2/z_O^3) + (\mu x_R^2/z_R^3) - (x_4^2/z_4^3). \tag{3.28}$$

For astigmatism to be eliminated, it is necessary to make $z_R = z_O$, $(x_P/z_P) = -(\mu x_R/z_R)$ and $z_P = \mu z_O$. The last condition is incompatible with the previously established condition for coma to vanish if $\mu \neq 1$, so that coma and astigmatism cannot be eliminated simultaneously unless $\mu = 1$.

3.4.4. Curvature of field

$$\begin{aligned} F = &[(x_P^2 + y_P^2)/z_P^3] - [\mu(x_O^2 + y_O^2)/z_O^3] \\ &+ [\mu(x_R^2 + y_R^2)/z_R^3] - [(x_4^2 + y_4^2)/z_4^3]. \end{aligned} \tag{3.29}$$

This coefficient also disappears when the astigmatism is reduced to zero.

3.4.5.   Distortion

$$D_x = [(x_P^3 + x_P y_P^2)/z_P^3 - [\mu(x_O^3 + x_O y_O^2)/z_O^3]$$
$$+ [\mu(x_R^3 + x_R y_R^2)/z_R^3] - [(x_4^3 + x_4 y_4^2)/z_4^3]. \tag{3.30}$$

Distortion cannot normally be eliminated when $\mu \neq 1$.

3.4.6.   Longitudinal distortion

It is apparent from (3.23) and (3.24) that unless $M_{lat} = 1$, and $\mu = 1$, the longitudinal magnification is not, in general, the same as the lateral magnification, resulting in a distortion in depth. This can be minimized for the case when the recording and reconstruction wavelengths are not the same ($\mu \neq 1$) by a proper choice of the recording and reconstruction geometry [Hariharan, 1976a].

3.4.7.   Nonparaxial imaging

The geometrical relations between the original object and the reconstructed images have been discussed in more detail by Neumann [1966], who describes a graphical method for locating the conjugate image. Depending on the conditions of illumination, the hologram may give two virtual images, two real images, or one virtual image and one real image. He has also shown that under some conditions only one image may exist, the energy of the second image wave then forming an evanescent field along the hologram.

The case of nonparaxial imaging was initially studied by Champagne [1967], who obtained expressions which can be used in many situations where paraxial theory is unsatisfactory. The possibility of using computer-based ray tracing methods based on these expressions was first investigated by Latta [1971a, b, c].

More precise imaging formulas that do not involve approximations for the inclination of the reference wave in recording or reconstruction have been derived by Miles [1972]. His results show that where the reference wave used in reconstruction differs appreciably from that used to record the hologram, the magnification in different azimuths is not, in general, the same, resulting in an anamorphic image. These formulas also permit precise calculations of the wavefront aberration by a method based on exact ray tracing [Miles, 1973].

3.5.   Effects of source size and spectral bandwidth

As described in the previous section, an image with no aberrations and the highest possible resolution can be obtained when the same coherent source used to record the hologram is also used to illuminate it. However, in many cases this is not feasible. We shall examine, in this section, how the use of a source of finite size and spectral bandwidth to illuminate the hologram affects the reconstructed image. For simplicity, we shall only consider the effects in the xz plane.

3.5.1.   Source size

Consider a hologram illuminated by a monochromatic point source of the same wavelength as that used to make it. From (3.9), the x coordinate of the virtual image of an object point located at (x_0, y_0, z_0) is

$$x_3 = \frac{x_P z_0 z_R + x_0 z_P z_R - x_R z_P z_0}{z_0 z_R + z_P z_R - z_P z_0}. \tag{3.31}$$

The displacement of the image for a small lateral shift in the position of the source is then given by the relation

$$(dx_3/dx_P) = \frac{z_0 z_R}{z_0 z_R + z_P z_R - z_P z_0}. \tag{3.32}$$

Accordingly, if the source used to illuminate the hologram is very nearly in the same position as the reference source used to record it ($z_P = z_R$), the image blur for a source size Δx_P can be written as

$$\Delta x_3 = (z_0/z_P)\Delta x_P. \tag{3.33}$$

The acceptable value of the image blur for a display is determined by the resolution of the eye, which is about 0.5 mrad, or 0.5 mm at a viewing distance of 1 m. Hence, if the image is located at a distance of 100 mm from the hologram, it can be illuminated with a spatially incoherent source such as a mercury vapour lamp with a diameter of 5 mm, provided the source is at a distance greater than 1 m from the hologram.

3.5.2.   Spectral bandwidth of the source

To calculate the effect of the spectral bandwidth of the source on the image, it is convenient to assume that the hologram is recorded with a plane reference wave using light of wavelength λ_1, but another plane reference wave with a wavelength λ_2, incident at the same angle, is used to illuminate the hologram when viewing the reconstructed image. Under these conditions, $z_P = z_R = \infty$ while the quantities $(x_P/z_P) = (x_R/z_R)$ are still finite, and the expressions, (3.9) and (3.11), derived earlier for the coordinates of an image point can be simplified to

$$x_3(\lambda_1, \lambda_2) = x_O + (x_P/z_P)(z_O/\mu) - (x_R/z_R)z_O,\tag{3.34}$$

and

$$z_3(\lambda_1, \lambda_2) = z_O/\mu,\tag{3.35}$$

where, as before, $\mu = \lambda_2/\lambda_1$.

The displacements of the image for a small change in the wavelength of the source used to illuminate the hologram are then given by the relations

$$(\mathrm{d}x_3/\mathrm{d}\lambda_2) = -(x_P/z_P)(z_O/\mu)(1/\lambda_2),\tag{3.36}$$

and

$$(\mathrm{d}z_3/\mathrm{d}\lambda_2) = -(z_O/\mu)(1/\lambda_2).\tag{3.37}$$

Hence, if the source used to illuminate the hologram has a mean wavelength λ_2 approximately equal to λ_1, so that $\mu \approx 1$, and a spectral bandwidth $\Delta\lambda_2$, the transverse image blur due to the finite spectral bandwidth of the source is

$$|\Delta x_3| = (x_P/z_P)z_O(\Delta\lambda_2/\lambda_2),\tag{3.38}$$

while the longitudinal image blur is

$$|\Delta z_3| = z_O(\Delta\lambda_2/\lambda_2).\tag{3.39}$$

While the longitudinal image blur is greater in magnitude, it is the transverse image blur which is usually more noticeable and which sets a limit on the spectral bandwidth of the source which can be used to illuminate a hologram. The permissible spectral bandwidth decreases if either the depth of the image or the interbeam angle is increased.

Typically, if green light from a high-pressure mercury vapour lamp, which has a mean wavelength λ_2 of 546 nm and a bandwidth of about 5 nm, is used to illuminate a hologram made with an argon-ion (Ar$^+$) laser ($\lambda_1 = 514$ nm) and an interbeam angle of 30° ($x_P/z_P = \tan 30° = 1/\sqrt{3}$), it can be seen from (3.38) that the transverse image blur $|\Delta x_3|$ will be equal to 0.5 mm when $z_O = 95$ mm.

3.5.3.   The image hologram

If the central plane of the image lies in the hologram plane, as in an image hologram (see section 2.5), the coherence requirements for the source used to illuminate the hologram are minimized. In fact, if the interbeam angle and the depth of the image are both made small, it is even possible to use an extended white-light source to illuminate the hologram. Points on the image in the hologram plane ($z_O = 0$) are then quite sharp and free from colour; other points in the image exhibit increasing colour dispersion and blur as their distance from this plane increases.

3.6. Image luminance

The luminance of the image reconstructed by a hologram depends primarily on its diffraction efficiency. This can be defined as the ratio of the energy diffracted into the desired image by an element of the hologram to that incident on it from the source used to illuminate it. The diffraction efficiency of a hologram is determined essentially by the recording medium used and the visibility of the carrier fringes. However, it can be shown that the luminance of the image also depends on the recording and reconstruction geometry [Hariharan, 1978].

3.6.1. The off-axis hologram

Consider a conventional off-axis hologram which, as shown in fig. 3.5, reconstructs a virtual image at a distance d behind it. This image is viewed by an observer located in front of the hologram.

Since the flux from a diffusely reflecting object incident on the photographic plate when the hologram was recorded can be assumed to be very nearly uniform over its whole area A_H, the visibility of the carrier fringes and, hence, the diffraction efficiency ε can be assumed to be constant over the hologram. Hence, if the hologram is illuminated by a monochromatic beam of intensity I and wavelength λ, the total energy diffracted into the image is $\varepsilon I A_H$.

Since, as can be seen from fig. 3.5, the flux from any element of the

Fig. 3.5. Ray paths in the reconstruction setup when the hologram itself is the pupil [Hariharan, 1978].

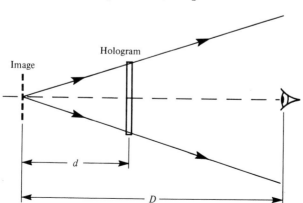

reconstructed image is spread out over a solid angle $\Omega_H = A_H/d^2$, the luminance of the image is

$$L_v = \varepsilon I A_H (K_\lambda/\Omega_H A_I),$$
$$= \varepsilon I K_\lambda d^2/A_I, \qquad (3.40)$$

where K_λ is the spectral luminous efficacy of the radiation and A_I is the area of the image.

It is apparent from (3.40) that the luminance of the image increases with its distance from the hologram; however this is at the expense of the available solid angle of viewing.

3.6.2.   The image hologram

It is not often realized that when a hologram is recorded of the real image projected either by an optical system or by another hologram (see section 3.3.1), it reconstructs an image not only of the object but also of the optical system including any aperture (or pupil) that limits the angular spread of the object beam. If the source used to illuminate the hologram has the same wavelength and occupies the same position with respect to the hologram as the reference source used to make it, the location and size of this reconstructed pupil correspond to those of the pupil of the imaging system or the boundaries of the primary hologram. As shown in fig. 3.6, the flux from any element of the reconstructed image is then confined within a

Fig. 3.6. Ray paths with an external reconstructed pupil [Hariharan, 1978].

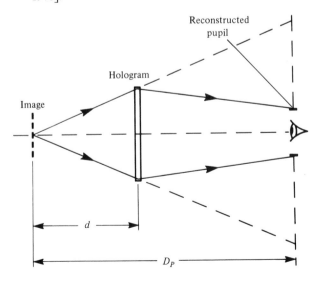

solid angle $\Omega_p = A_p/D_p^2$, where A_p is the area of the reconstructed pupil and D_p is its distance from the image.

If the image is located at an appreciable distance from the hologram, and the numerical aperture of the imaging system used to make the hologram is large enough, the flux in the object beam can be assumed to be very nearly uniform, and the diffraction efficiency of the hologram will be nearly constant over its whole area. The luminance of the image is then

$$L_v = \varepsilon I A_H (K_\lambda/\Omega_p A_I).$$

$$(3.41)$$

For a given hologram size and viewing distance, the luminance of the image is now independent of its distance from the hologram and depends only on the dimensions of the reconstructed pupil. A comparison of (3.41) and (3.40) shows that the luminance of the image has increased by a factor (Ω_H/Ω_p), which is the reciprocal of the ratio of the solid angles of viewing available in the two cases. A substantial improvement in image luminance is possible by the formation of an external pupil whose shape and size match the range of angles over which the hologram is actually to be viewed.

As described in section 3.5.3 a major advantage of image holograms is that the image can also be made to occupy a position in the hologram plane: this minimizes the image blur when a source of finite size and finite spectral bandwidth is used to illuminate the hologram. In this case only the area of the hologram corresponding to the image diffracts light and (3.41) reduces to

$$L_v = \varepsilon I K_\lambda/\Omega_p,$$

$$(3.42)$$

so that the luminance of the image is independent of its area. However, it is apparent from (3.41) and (3.42) that the image luminance in this case must be less than the luminance of an image of the same size formed at a distance from the hologram. To maximize image luminance, the object should be at a sufficient distance so that, while recording the hologram, the flux from the object is spread out over the photographic plate. This distance is, of course, limited by the size of the source to be used to illuminate the hologram and the acceptable image blur.

3.7. Image speckle

When a diffusely reflecting object is illuminated by a coherent source, each of the microscopic elements making up the surface of the object gives rise to a coherent diffracted wave. If then, as is usually the case, the point-spread function of the imaging system is broad compared to the dimensions of these microscopic elements, the diffracted amplitudes from many elements will contribute to the resultant amplitude at each point in the reconstructed image. Since the scale of the surface irregularities is such that the optical paths to neighbouring elements exhibit random differences which may amount to several wavelengths, the interference of these diffracted components gives rise to local fluctuations in the image intensity (see Appendix A4). As a result the image exhibits a granular or speckled appearance.

Speckle is a serious problem in holographic imaging. A number of methods have been described to reduce speckle in the reconstructed image (see, for example, the detailed review by McKechnie [1975c]), and some of these will be discussed here.

If the subject is a two-dimensional transparency, speckle does not appear when a uniform plane or spherical wave is used to illuminate it. However, a major drawback in this case is that any dust particles or imperfections on the surface of the transparency give rise to annoying diffraction patterns in the reconstructed image. These can be minimized with a Fourier transform recording geometry, but then most of the power in the object beam is concentrated in a small spot. This results in poor diffraction efficiency and can lead to nonlinearity (see section 6.5), unless a relatively intense reference beam is used.

These problems can be avoided by illuminating the transparency through a diffuser when recording the hologram. A further advantage then is that each point on the hologram receives light from every point of the transparency, so that the information stored on the hologram is no longer localized. As a result, damage to the hologram only results in a slight increase in scattered light, unlike in a photograph where it can result in complete loss of information at that particular point. A spectacular consequence is that even if the hologram is broken, any fragment will reconstruct the complete image, though with reduced resolution. However, the penalty paid for this is that the image appears modulated on the random speckle pattern due to the diffuser.

If, however, the point-spread function of the imaging system is narrow enough to resolve individual elements on the diffuser, the complex amplitude at the diffuser is reproduced perfectly in the image plane. The image then exhibits a pure phase modulation and no speckle is observed. The same result is obtained if the angular spread of light from the diffuser is made small enough for all the light from it to pass through the aperture of the imaging system.

It is possible, therefore, to minimize speckle while retaining most of the advantages of diffuse illumination, by using, instead of a random diffuser, a quasi-random phase plate [Upatnieks, 1967] or phase gratings [Gerritsen, Hannan & Ramberg, 1968; Gabor, 1970; Kato & Okino, 1973; Matsumura, 1975] in contact with the transparency.

Unfortunately, this technique can be used only with nondiffusing objects and then only for a single plane in the object. Accordingly, a number of methods of speckle reduction have been described for diffusing objects. All of them have some disadvantages, and the choice is usually determined by the sacrifices that can be made.

3.7.1. Low-pass filtering

The simplest method of speckle reduction is merely to form an image using a hologram (or lens) aperture large enough that the average size of the speckles is significantly smaller than the resolution limit of the photographic material used to record the image. In this case, the exposure at any point is determined by the intensity in the image averaged over a number of speckles, so that the intensity fluctuations are reduced significantly.

3.7.2. Partially coherent illumination

Since speckle is a consequence of the coherence of the illumination, it can be reduced by using partially coherent illumination to reconstruct the image. With an image hologram it is possible to use a white-light source [Golbach, 1973]. Alternatively, the spatial coherence of the beam can be reduced by means of moving diffusers [Lowenthal & Joyeux, 1971; Ih & Baxter, 1978].

3.7.3. Redundancy

In one method [Martienssen & Spiller, 1967] the object is illuminated through a diffuser and a number of holograms are made with the diffuser in different positions. If the images reconstructed by N such holograms are superposed on the same photographic film by successive exposures, each exposure records the same information on the object but a different speckle pattern. This results in a reduction in the contrast of the speckle pattern by a factor of $N^{1/2}$. An alternative method which gives the same result is the use of wavelength diversity [George & Jain, 1973].

3.7.4. Time-averaging

Instead of superposing the images from a number of holograms, as described in section 3.7.3, it is possible to sample a single large hologram by means of a small moving pupil [Dainty & Welford, 1971]. As the pupil uncovers different parts of the hologram, the speckle pattern in the image plane changes so that the exposure corresponds to the sum of the irradiance distributions in a number of uncorrelated speckle patterns. This method was extended by Yu & Wang [1973], who proposed the use of a moving mask with a random distribution of openings covering the entire hologram, and by Som & Budhiraja [1975], who claimed that a further reduction in speckle could be obtained by the use of a moving mask at some distance from the hologram, both in recording and reconstruction. However, it can be shown that while there is a reduction of speckle, it is always accompanied by a reduction in either the resolution or the contrast of the image [Hariharan & Hegedus, 1974a, b; McKechnie, 1975a, b; Östlund & Biedermann, 1977].

4

Types of holograms

A hologram recorded on a photographic plate and processed normally is equivalent to a grating with a spatially varying transmittance. However, with suitable processing, it is possible to produce a spatially varying phase shift. In addition, if the thickness of the recording medium is large compared with the fringe spacing, volume effects are important. In an extreme case, it is even possible to produce holograms in which the fringes are planes running almost parallel to the surface of the recording material and which reconstruct an image in reflected light.

Based on these characteristics, holograms recorded in a thin recording medium can be divided into amplitude holograms and phase holograms. Holograms recorded in relatively thick recording media can be classified either as transmission amplitude holograms, transmission phase holograms, reflection amplitude holograms or reflection phase holograms.

In the next few sections we shall examine some of the principal characteristics of these six types of holograms.

4.1. Thin holograms

Any hologram in which the thickness of the recording material is small compared with the average spacing of the interference fringes can be classified as a thin hologram. Such a hologram can be characterized by a spatially varying complex amplitude transmittance

$$\mathbf{t}(x, y) = |\mathbf{t}(x, y)| \exp\left[-i\phi(x, y)\right]. \tag{4.1}$$

4.1.1. Thin amplitude holograms

In an amplitude hologram, $\phi(x, y)$ is essentially constant while $|\mathbf{t}(x, y)|$ varies over the hologram. To calculate the complex amplitude of the diffracted waves from such a hologram and, hence, its diffraction efficiency, consider a grating formed in a suitable thin recording medium by a plane object wave and a plane reference wave.

If we assume that the resulting amplitude transmittance is linearly related to the intensity in the interference pattern, the amplitude transmittance of the grating can be written as

$$|\mathbf{t}(x)| = \mathbf{t}_0 + \mathbf{t}_1 \cos Kx. \tag{4.2}$$

where $\mathbf{t}_0$ is the average amplitude transmittance of the grating, $\mathbf{t}_1$ is the amplitude of the spatial variation of $|\mathbf{t}(x)|$ and

$$K = 2\pi/\Lambda, \tag{4.3}$$

where Λ is the spacing of the fringes.

Since the values of $|\mathbf{t}(x)|$ are limited to the range $0 \leqslant |\mathbf{t}(x)| \leqslant 1$, and the amplitudes of the diffracted waves are linearly proportional to the amplitude of the spatial variation of $|\mathbf{t}(x)|$, the diffracted amplitude is a maximum when

$$\begin{aligned} |\mathbf{t}(x)| &= (1/2) + (1/2) \cos Kx, \\ &= (1/2) + (1/4) \exp(iKx) + (1/4) \exp(-iKx). \end{aligned}$$
$$\tag{4.4}$$

The maximum amplitude in each of the diffracted orders is one fourth of that in the wave used to illuminate the hologram, so that the peak diffraction efficiency is

$$\varepsilon_{\text{max}} = (1/16), \tag{4.5}$$

or 0.0625.

In practice, no recording medium has a linear response over the full range of transmittance values from 0 to 1; hence this value of ε cannot be achieved without running into nonlinear effects.

4.1.2.   Thin phase holograms

For a lossless phase grating, $|\mathbf{t}(x)| = 1$, so that the complex amplitude transmittance is

$$\mathbf{t}(x) = \exp[-i\phi(x)]. \tag{4.6}$$

If the phase shift produced by the recording medium is linearly proportional to the intensity in the interference pattern,

$$\phi(x) = \phi_0 + \phi_1 \cos(Kx), \tag{4.7}$$

and the complex amplitude transmittance of the grating is

$$\mathbf{t}(x) = \exp(-i\phi_0) \exp[-i\phi_1 \cos(Kx)]. \tag{4.8}$$

If we neglect the constant phase factor $\exp(-i\phi_0)$, the right hand side of (4.8) can be expanded as a Fourier series to give

$$\mathbf{t}(x) = \sum_{n=-\infty}^{\infty} i^n J_n(\phi_1) \exp(inKx), \tag{4.9}$$

where J_n is the Bessel function of the first kind of order n.

Such a thin phase grating diffracts a wave incident on it into a large number of orders, the diffracted amplitude in the nth order being proportional to the value of the Bessel function $J_n(\phi_1)$. Only the wave of order 1 contributes to the desired image. As shown in fig. 4.1, the amplitude diffracted into this order, which is proportional to $J_1(\phi_1)$, increases at first with the phase modulation and then decreases.

The diffraction efficiency of the grating is, accordingly,

$$\varepsilon = J_1^2(\phi_1),$$ (4.10)

and its maximum value is

$$\varepsilon_{max} = 0.339.$$ (4.11)

4.2. Volume holograms

The medium in which a hologram is recorded can have a thickness of as much as a few millimetres, while the fringe spacing may only be of the order of 1 μm. The hologram is then a three-dimensional system of layers corresponding to a periodic variation of absorption or refractive index, and the diffracted amplitude is a maximum only when the Bragg condition is satisfied. The characteristics of generalized volume gratings have been discussed in detail by Russell [1981] and by Solymar & Cooke [1981]. However, for simplicity, we will only consider a grating produced by recording the interference of two infinite plane wavefronts in a thick

Fig. 4.1. Diffraction efficiency of a thin phase grating as a function of the phase modulation [Kogelnik, 1967].

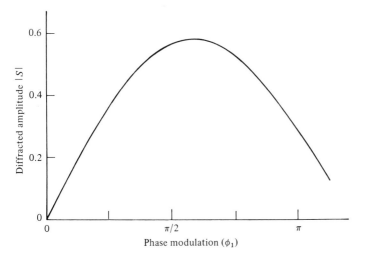

Phase modulation (ϕ_1)

recording medium. We will also assume that, initially, the recording medium is perfectly transparent but, after processing, develops a sinusoidal variation of the absorption or the refractive index in the direction perpendicular to the interference surfaces. In addition, while the interference surfaces can assume any orientation, only two limiting cases will be considered, in which they are either perpendicular or parallel to the hologram plane.

The first case arises when the two interfering wavefronts make equal but opposite angles to the surface of the recording medium and are incident on it from the same side. Holograms recorded in this fashion give a reconstructed image in transmitted light. The second case arises when the wavefronts are symmetrical with respect to the surface of the recording medium but are incident on it from opposite sides. Holograms of this type give a reconstructed image by reflection.

The spacing between successive fringe planes is a minimum, and the volume effects are most pronounced, when the angle between the two interfering wavefronts is a maximum ($\approx 180°$). This has made it possible to produce reflection holograms with a wavelength selectivity high enough to reconstruct an image of acceptable quality when illuminated with white light.

4.3.   The coupled wave theory

When analysing the diffraction of light by such thick gratings, it is necessary to take into account the fact that the amplitude of the diffracted wave increases progressively, while that of the incident wave decreases, as they propagate through the grating. One way of doing this is by means of a coupled wave approach such as that developed by Kogelnik [1967, 1969].

Consider a coordinate system in which, as shown in fig. 4.2, the z-axis is perpendicular to the surfaces of the recording medium and the x-axis is in the plane of incidence, while the fringe planes are oriented perpendicular to the plane of incidence. The grating vector $\mathbf{K}$ is perpendicular to the fringe planes. It is of length $|\mathbf{K}| = 2\pi/\Lambda$, where Λ is the grating period, and makes an angle ψ ($\psi = 90°$ or $0°$ in the cases shown) with the z axis. The refractive index n and the absorption constant α are assumed to vary sinusoidally, their values at any point being given by the relations

$$n = n_0 + n_1 \cos \mathbf{K} \cdot \mathbf{x}, \tag{4.12}$$

$$\alpha = \alpha_0 + \alpha_1 \cos \mathbf{K} \cdot \mathbf{x}, \tag{4.13}$$

where the radius vector $\mathbf{x} = (x, y, z)$. For simplicity, the refractive index of the surrounding medium is also assumed to be n_0.

If monochromatic light is incident on the hologram grating at, or near, the Bragg angle, and if the thickness of the medium is large enough, only two waves in the grating need be taken into consideration; these are the incoming reference wave R and the outgoing signal wave S. Since the other diffraction orders violate the Bragg condition strongly, they are severely attenuated and can be neglected. If we also assume that these waves are polarized with their electric vector perpendicular to the plane of incidence, their interaction in the grating can be described by the scalar wave equation

$$\nabla^2 E + k^2 E = 0, \tag{4.14}$$

where E is the total electric field and k is the (spatially varying) propagation constant in the grating.

We assume that the absorption per wavelength as well as the relative variations in refractive index of the medium are small, so that

$$n_0 k_0 \gg \alpha_0,$$
$$n_0 k_0 \gg \alpha_1,$$
$$n_0 \gg n_1, \tag{4.15}$$

Fig. 4.2. Volume transmission and reflection gratings and their associated vector diagrams for Bragg incidence [Kogelnik, 1967].

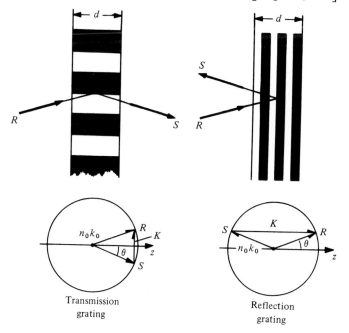

Transmission grating

Reflection grating

where $k_0 = 2\pi/\lambda$. The propagation constant can then be written in the form

$$k^2 = B^2 - 2i\alpha_0 B + 4\kappa B \cos \mathbf{K} \cdot \mathbf{x}, \tag{4.16}$$

where $B = n_0 k_0$ is the average propagation constant, and κ is the coupling constant defined as

$$\kappa = (\pi n_1/\lambda) - i\alpha_1/2. \tag{4.17}$$

This coupling constant describes the interaction between the reference wave R and the signal wave S. If $\kappa = 0$, there is no modulation of the refractive index or the absorption and hence, no diffraction.

The propagation of the two coupled waves through the grating can be described by their complex amplitudes $R(z)$ and $S(z)$, which vary along z as a result of the energy interchange between them as well as energy losses due to absorption. The total electric field E in the grating is then the superposition of the fields due to these two waves so that

$$E = R(z) \exp(-i\boldsymbol{\rho} \cdot \mathbf{x}) + S(z) \exp(-i\boldsymbol{\sigma} \cdot \mathbf{x}), \tag{4.18}$$

where $\boldsymbol{\rho}$ and $\boldsymbol{\sigma}$ are the propagation vectors for the two waves; these are defined by the propagation constants and the directions of propagation of R and S. The quantity $\boldsymbol{\rho}$ is assumed to be equal to the propagation vector of the free reference wave in the absence of coupling, while $\boldsymbol{\sigma}$ is determined by the grating and is related to $\boldsymbol{\rho}$ and the grating vector by the expression

$$\boldsymbol{\sigma} = \boldsymbol{\rho} - \mathbf{K}. \tag{4.19}$$

For the special case of incidence at the Bragg angle θ_0, the lengths of both $\boldsymbol{\rho}$ and $\boldsymbol{\sigma}$ are equal to the free propagation constant $n_0 k_0$, and the Bragg condition, which can be written in the form

$$\cos(\psi - \theta_0) = K/2n_0 k_0, \tag{4.20}$$

is obeyed. If (4.20) is differentiated, we obtain

$$(\mathrm{d}\theta_0/\mathrm{d}\lambda_0) = K/4\pi n_0 \sin(\psi - \theta_0). \tag{4.21}$$

It follows from (4.21) that small changes in the angle of incidence or the wavelength have similar effects.

A useful parameter for evaluating the effects of deviations from the Bragg condition is the dephasing measure ζ, which can be defined as

$$\begin{aligned}
\zeta &= (|\boldsymbol{\rho}|^2 - |\boldsymbol{\sigma}|^2)/2|\boldsymbol{\rho}|, \\
&= (B^2 - |\boldsymbol{\sigma}|^2)/2B, \\
&= K \cos(\psi - \theta) - K^2\lambda/4\pi n_0,
\end{aligned} \tag{4.22}$$

from (4.19). For small deviations $\Delta\theta$ and $\Delta\lambda$ from the Bragg condition, this becomes

$$\zeta = \Delta\theta \cdot K \sin(\psi - \theta_0) - \Delta\lambda \cdot K^2/4\pi n_0. \tag{4.23}$$

To derive the coupled wave equations, (4.14) and (4.16) are combined, and (4.18) and (4.19) are inserted. If then, the terms involving $\exp(-i\boldsymbol{\rho}\cdot\mathbf{x})$ and $\exp(-i\boldsymbol{\sigma}\cdot\mathbf{x})$ are compared, we get

$$R'' - 2iR'\rho_z - 2i\alpha BR + 2\kappa BS = 0, \tag{4.24}$$

and

$$S'' - 2iS'\sigma_z - 2i\alpha BS + (B^2 - |\boldsymbol{\sigma}|^2)S + 2\kappa BR = 0, \tag{4.25}$$

where the primes denote differentiation with respect to z. If, in addition, it is assumed that the energy interchange between S and R, as well as the energy absorption in the medium are slow, the second differentials R'' and S'' can be neglected. From (4.23) these equations can then be rewritten in the form

$$R'\cos\theta + \alpha R = i\kappa S, \tag{4.26}$$

$$[\cos\theta - (K/B)\cos\psi]S' + (\alpha + i\zeta)S = -i\kappa R. \tag{4.27}$$

The coupled wave equations (4.26) and (4.27) show that the amplitude of a wave changes along z because of coupling to the other wave (κR, κS) or absorption (αR, αS). For deviations from the Bragg condition, S is forced out of synchronism with R, due to the term involving ζS, and the interaction decreases.

The coupled wave equations, (4.26) and (4.27), can be solved for the appropriate boundary conditions. These are $R(0) = 1$, $S(0) = 1$, for transmission gratings, and $R(0) = 1$, $S(d) = 0$, for reflection gratings.

In the next few sections we will discuss the solutions for the most important cases, namely lossless phase gratings and pure absorption gratings, the grating planes being assumed to run either normal to the surface (for transmission gratings) or parallel to the surface (for reflection gratings). The method of solution of the coupled wave equations, as well as solutions for the cases of slanted gratings, lossy phase gratings and mixed gratings are to be found in the original paper by Kogelnik [1969]. This paper also gives an extension of the theory to light polarized with the electric vector in the plane of incidence.

4.4. Volume transmission holograms

4.4.1. Phase gratings

In a lossless phase grating $\alpha_0 = \alpha_1 = 0$. Diffraction is caused by the spatial variation of the refractive index. The diffracted amplitude is then

$$S(d) = \frac{-i\exp(-i\chi)\sin(\Phi^2 + \chi^2)^{1/2}}{(1 + \chi^2/\Phi^2)^{1/2}}, \tag{4.28}$$

where

$$\Phi = \pi n_1 d / \lambda \cos \theta, \tag{4.29}$$

and

$$\chi = \zeta d / 2 \cos \theta. \tag{4.30}$$

From (4.23) it follows that for incidence at the Bragg angle, $\zeta = 0$, so that $\chi = 0$ and (4.28) becomes

$$S(d) = -i \sin \Phi. \tag{4.31}$$

Since the incident amplitude $R(0)$ is assumed to be unity, the diffraction efficiency is

$$\varepsilon = |S(d)|^2 = \sin^2 \Phi. \tag{4.32}$$

As either d the thickness, or n_1 the variation of the refractive index, increases, the diffraction efficiency increases until the modulation parameter $\Phi = \pi/2$. At this point $\varepsilon = 1.00$, and all the energy is in the diffracted beam. When Φ increases beyond this point, energy is coupled back into the incident wave, and ε drops.

When the angle of incidence or the wavelength of the incident beam deviates from the values required to satisfy the Bragg condition, the diffraction efficiency drops to

$$\varepsilon = \frac{\sin^2(\Phi^2 + \chi^2)^{1/2}}{(1 + \chi^2/\Phi^2)}. \tag{4.33}$$

The effect of angular and wavelength deviations $\Delta\theta$ and $\Delta\lambda$ from the Bragg condition can be studied readily, since they influence the diffraction efficiency mainly through the parameter χ, which is a measure of the deviation from the Bragg condition. From (4.23) it can be rewritten in the form

$$\chi = \Delta\theta \cdot K d/2, \tag{4.34}$$

or, alternatively

$$\chi = -\Delta\lambda \cdot K^2 d / 8\pi n_0 \cos \theta_0, \tag{4.35}$$

while the modulation parameter Φ can be taken as constant.

Curves showing the normalized diffraction efficiency of a lossless transmission phase hologram as a function of the parameter χ are plotted in fig. 4.3 for three values of the modulation parameter Φ. For a hologram with $\Phi = \pi/2$, which diffracts with maximum efficiency (1.00) at the Bragg angle, the diffraction efficiency drops to zero when $\chi = 2.7$.

4.4.2. Effect of loss

For a lossy phase grating (absorption constant α_0), the diffracted amplitude at the Bragg angle is

$$S(d) = -i \exp(-\alpha_0 d/\cos\theta) \sin\Phi. \tag{4.36}$$

This expression has the same form as (4.31), except for an additional exponential term containing the absorption coefficient $\alpha_0 d$. This term mainly decreases the peak diffraction efficiency and has only a small effect on the angular and wavelength selectivity.

4.4.3. Amplitude gratings

In an amplitude grating, the refractive index does not vary, so that $n_1 = 0$. However, the absorption constant varies with an amplitude α_1 about its mean value α_0. In this case, the coupling constant $\kappa = -i\alpha_1/2$, and the diffracted amplitude is

Fig. 4.3. Volume transmission holograms. Angular and wavelength sensitivity of a phase grating showing the normalized diffraction efficiency $(\varepsilon/\varepsilon_0)$ as a function of the parameter χ, which is a measure of the deviation from the Bragg condition, for three different values of the modulation parameter Φ [Kogelnik, 1969].

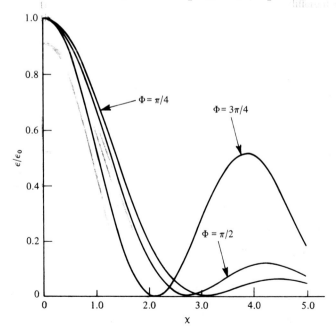

$$S(d) = -\exp\left(-\frac{\alpha_0 d}{\cos\theta}\right)\exp(-i\chi)\frac{\sinh(\Phi_a^2 - \chi^2)^{1/2}}{(1 - \chi^2/\Phi_a^2)^{1/2}}, \tag{4.37}$$

where $\Phi_a = \alpha_1 d/2\cos\theta$, and χ is defined, as before, by (4.30).

For incidence at the Bragg angle, $\chi = 0$, and the diffraction efficiency can be written as

$$\varepsilon = \exp(-2\alpha_0 d/\cos\theta_0)\sinh^2(\alpha_1 d/2\cos\theta_0). \tag{4.38}$$

The diffracted amplitude increases with α_1, but since negative values of the absorption are excluded, $\alpha_1 \leqslant \alpha_0$. The highest diffraction efficiency is therefore obtained when $\alpha_1 = \alpha_0$, for a value of $\alpha_1 d\cos\theta_0 = \ln 3$; this maximum has a value

$$\varepsilon_{\max} = 1/27, \tag{4.39}$$

or 0.037.

Figure 4.4 shows the diffracted amplitude computed from (4.38) as a function of two parameters, $D_0 = \alpha_0 d/\cos\theta_0$ and $D_1 = \alpha_1 d/\cos\theta_0$. The

Fig. 4.4. Volume transmission holograms. Amplitude diffracted by an amplitude grating as a function of the modulation $D_1 = \alpha_1 d/\cos\theta_0 = 2\Phi_a$, for various bias levels $D_0 = \alpha_0 d/\cos\theta_0$ (broken lines) and modulation depths D_1/D_0 (solid curves) [Kogelnik, 1969].

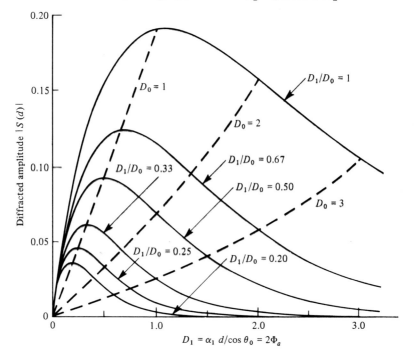

quantity D_1 is a measure of the amplitude of the modulation, while $D_1/D_0 = \alpha_1/\alpha_0$ is a measure of the relative depth of modulation. The curves for constant D_0 show the behaviour for a constant background absorption and varying relative depth of modulation. A linear response and relatively good efficiency is obtained for a background absorption $D_0 = 1$.

Curves showing the angular and wavelength sensitivity of amplitude gratings computed from (4.37) are presented in fig. 4.5. These curves are plotted as functions of the parameter χ, which is a measure of the deviation from the Bragg condition, for the special case of $\alpha_1 = \alpha_0$ and for values of the modulation parameter $\Phi_a = \alpha_1 d/2 \cos \theta$ of 0.55 (corresponding to the peak diffraction efficiency of 0.037) and 1.0. The curves are similar and, in both cases, the diffraction efficiency drops to zero when $\chi \approx 3.3$.

4.5. Volume reflection holograms

As shown in fig. 4.2, the fringe planes in a reflection hologram run more or less parallel to the surface of the recording medium so that $\psi = 0$.

Fig. 4.5. Volume transmission holograms. Angular and wavelength sensitivity of an amplitude grating for $\alpha_1 = \alpha_0 (D_1 = D_0)$ and values of $\Phi_a = D_1/2 = 0.55 (\varepsilon_0 = 0.037)$ and $\Phi_a = D_1/2 = 1 (\varepsilon_0 = 0.025)$. The curves show the normalized diffraction efficiency $(\varepsilon/\varepsilon_0)$ as a function of the parameter χ which is a measure of the deviation from the Bragg condition [Kogelnik, 1969].

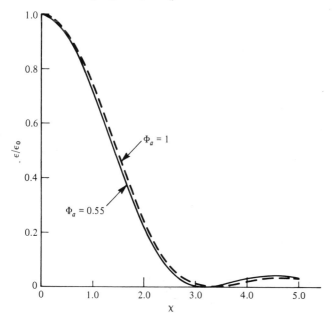

When the hologram is illuminated by a wave incident from the left, the diffracted wave starts with an amplitude $S(d) = 0$ at the rear face of the hologram, but gains in amplitude as it propagates towards the left.

4.5.1. Phase gratings

In a lossless phase grating $\alpha_0 = \alpha_1 = 0$, and the coupling constant $\kappa = \pi n_1/\lambda$. For small deviations from the Bragg angle the diffracted amplitude is given by the expression

$$S(0) = i[(i\chi_r/\Phi_r) + (1 - \chi_r^2/\Phi_r^2)^{1/2} \coth (\Phi_r^2 - \chi_r^2)^{1/2}]^{-1}, \tag{4.40}$$

where

$$\Phi_r = \pi n_1 d/\lambda \cos \theta, \tag{4.41}$$

and

$$\chi_r = \zeta d/2 \cos \theta. \tag{4.42}$$

The diffraction efficiency is then

$$\varepsilon = [1 + (1 - \chi_r^2/\Phi_r^2)/\sinh^2 (\phi_r^2 - \chi_r^2)^{1/2}]^{-1}. \tag{4.43}$$

For a wave incident at the Bragg angle, $\chi_r = 0$ and (4.43) can be simplified and written as

$$\varepsilon = \tanh^2 (\pi n_1 d/\lambda \cos \theta_0). \tag{4.44}$$

As shown in fig. 4.6, the diffraction efficiency increases asymptotically to 1.00 as the value of Φ_r increases.

Curves showing the angular sensitivity and the wavelength sensitivity of a lossless phase grating computed from (4.43) are presented in fig. 4.7 using the following relationships derived from (4.42) for the parameter χ_r, which is a measure of the deviation from the Bragg condition.

$$\begin{aligned}\chi_r &= \Delta\theta(2\pi n_0 d/\lambda) \sin \theta_0, \\ &= (\Delta\lambda/\lambda)(2\pi n_0 d/\lambda) \cos \theta_0.\end{aligned} \tag{4.45}$$

These curves are plotted for values of the modulation parameter Φ_r of $\pi/4$, $\pi/2$ and $3\pi/4$. While the diffraction efficiency drops to zero in all cases when $\chi_r \approx 3.5$, the selectivity curves become appreciably broader with increasing values of Φ_r.

4.5.2. Amplitude gratings

In this case, as in section 4.4.3, $n_1 = 0$, while α_0 and α_1 are finite. The amplitude of the diffracted wave leaving the hologram is then

$$S(0) = -\{(\chi_{ra}/\Phi_{ra}) + [(\chi_{ra}/\Phi_{ra})^2 - 1]^{1/2} \coth (\chi_{ra}^2 - \Phi_{ra}^2)^{1/2}\}^{-1}, \tag{4.46}$$

Fig. 4.6. Volume reflection holograms. Diffraction efficiency of a phase grating at the Bragg angle as a function of the modulation parameter Φ_r.

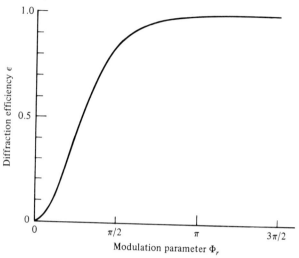

Fig. 4.7. Volume reflection holograms. Angular and wavelength sensitivity of a phase grating, showing the normalized diffraction efficiency, $(\varepsilon/\varepsilon_0)$ as a function of the parameter χ_r, which is a measure of the deviation from the Bragg condition, for different values of the modulation parameter Φ_r [Kogelnik, 1969].

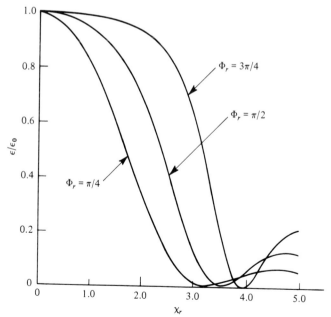

where

$$\Phi_{ra} = \alpha_1 d/2 \cos \theta, \qquad (4.47)$$

and

$$\chi_{ra} = (\alpha_0 d/\cos \theta_0) - (i\zeta d/2 \cos \theta_0). \qquad (4.48)$$

If the incident wave is at the Bragg angle, $\zeta = 0$ and $\chi_{ra}/\Phi_{ra} = 2\alpha_0/\alpha_1$, so that (4.46) can be written as

$$S(0) = -\{(2\alpha_0/\alpha_1) + [(4\alpha_0^2/\alpha_1^2) - 1]^{1/2} \coth [(d/\cos \theta_0)(\alpha_0^2 - \alpha_1^2/4)^{1/2}]\}^{-1}.$$

$$(4.49)$$

Curves of the diffracted amplitude as a function of the modulation parameter $D_1 = \alpha_1 d/\cos \theta_0 = 2\Phi_{ra}$ are presented in fig. 4.8 for different values of the modulation depth $D_1/D_0 = \alpha_1/\alpha_0$ and bias level $D_0 = \alpha_0 d/\cos \theta_0$. These show clearly that for reflection holograms of this type, the best

Fig. 4.8. Volume reflection holograms. Amplitude diffracted by an amplitude grating as a function of the modulation $D_1 = \alpha_1 d/\cos \theta_0 = 2\Phi_{ra}$ for various bias levels $D_0 = \alpha_0 d/\cos \theta_0$ (broken lines) and modulation depths D_1/D_0 (solid curves) [Kogelnik, 1969].

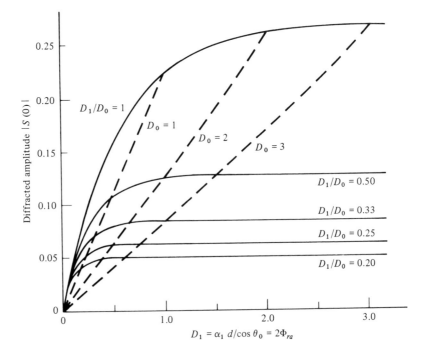

efficiency is obtained when $\alpha_0 d/\cos\theta_0 \geqslant 2$, corresponding to an optical density $\geqslant 1.7$. (See Appendix A5.)

For the special case when the modulation depth is a maximum ($\alpha_1 = \alpha_0$), (4.49) can be simplified further to

$$S(0) = -[2 + 3^{1/2}\coth(3^{1/2}\alpha_0 d/2\cos\theta_0)]^{-1}, \qquad (4.50)$$

from which it is apparent that the maximum possible value of the diffraction efficiency is

$$\varepsilon_{max} = (2 + \sqrt{3})^{-2}, \qquad (4.51)$$

or 0.072.

Figure 4.9 shows curves for the angular selectivity and wavelength selectivity of an amplitude grating derived from (4.46), making use of the fact that (4.48) can be rewritten in the form

$$\chi_{ra} = (\alpha_0 d/\cos\theta_0) - i\Delta\theta(2\pi n_0 d/\lambda_0)\sin\theta_0.$$
$$= (\alpha_0 d/\cos\theta_0) - i(\Delta\lambda/\lambda_0)(2\pi n_0 d/\lambda_0)\cos\theta_0. \qquad (4.52)$$

Fig. 4.9. Volume reflection holograms. Angular and wavelength sensitivity of an amplitude grating for $D_1 = D_0$ and values of $D_1 = 0.2$ ($\varepsilon_0 = 0.007$), $D_1 = 1.0$ ($\varepsilon_0 = 0.05$) and $D_1 = 2$ ($\varepsilon_0 = 0.068$). The normalized diffraction efficiency ($\varepsilon/\varepsilon_0$) is plotted as a function of the term $\zeta d/2\cos\theta_0$, which is a measure of the deviation from the Bragg condition [Kogelnik, 1969].

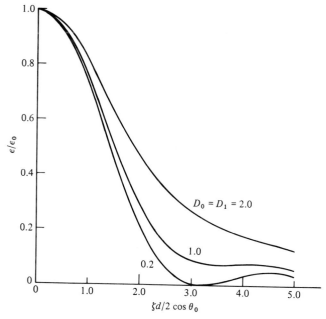

In these curves, the diffracted amplitude is plotted as a function of the term $\zeta d/2 \cos \theta_0$, which is a measure of the deviation from the Bragg condition, for three values of the modulation parameter D_1 when the modulation depth is a maximum $(\alpha_1/\alpha_0 = 1)$. The curves become broader as the loss parameter D_0 increases.

4.6.   Discussion

The maximum diffraction efficiencies that can be obtained with the six types of gratings studied are summarized in Table 4.1.

While the maximum diffraction efficiency that can actually be obtained with a grating is limited by nonlinearities in the response of the recording medium and consequent distortions of the profile of the fringes, values approximating these theoretical limits have been obtained in a number of experiments, confirming the conclusions of the coupled wave theory.

On the other hand, the maximum diffraction efficiency that can be obtained with a hologram of a diffusely reflecting object is always much lower, because the spatial variations in the amplitude of the object wave (see Appendix A4) make it impossible to maintain optimum modulation over the entire area.

The average diffraction efficiencies of transmission phase holograms recorded with a diffuse object beam in both thin and thick recording media have been calculated by Upatnieks & Leonard [1970] assuming that the amplitude of the object wave has a Rayleigh probability distribution. These calculations show that the maximum diffraction efficiency is 0.22 with a thin recording medium, and 0.64 with a thick recording medium, against 0.339 and 1.00, respectively, for phase gratings recorded in the same media.

Table 4.1. *Maximum theoretical diffraction efficiency for different types of gratings*

Type of grating	Thin transmission		Volume transmission		Volume reflection	
Modulation	Amplitude	Phase	Amplitude	Refractive index	Amplitude	Refractive index
ε_{max}	0.0625	0.339	0.037	1.00	0.072	1.00

4.7. More accurate theories

Kogelnik's coupled wave theory assumes infinite plane wavefronts as well as a slow energy interchange and only a small absorption loss per wavelength. Two other major assumptions are that the modulation is sinusoidal and only the zero-order and Bragg diffracted waves propagate within the grating. However, even if the first of these is satisfied, the existence of higher-order waves in holographic phase gratings cannot be ignored, even for fairly small values of the refractive index modulation ($n_1 \geqslant 0.005$).

More accurate theories for the diffraction of light by volume holograms were therefore developed by a number of authors, who followed two other basic approaches. These were a rigorous modal approach [Burckhardt, 1966a; Chu & Tamir, 1970; Kaspar, 1973] and a thin-grating decomposition approach [Alferness, 1975a, 1976]. The equivalence of these two approaches and the coupled wave theory has been discussed by Alferness [1975b], Magnusson & Gaylord [1978b] and Gaylord & Moharam [1982]. More general treatments based on the coupled wave theory have also been formulated by Su & Gaylord [1972], Kessler & Kowarschik [1975], Kowarschik & Kessler [1975], Magnusson & Gaylord [1977] and Moharam & Gaylord [1981]. These cover gratings with nonsinusoidal profiles and lead to equations which can be solved with the aid of a digital computer. As expected, they confirm that at larger values of the refractive index modulation a substantial amount of power can be diffracted into higher orders, and the diffraction efficiency at the second- and third-order Bragg angles can be quite high (> 0.2).

Effects due to the sharp boundaries between the sinusoidally modulated grating and the surrounding medium have also been ignored in Kogelnik's theory. These have been taken into consideration in a recent treatment using the modal theory [Langbein & Lederer, 1980; Lederer & Langbein, 1980], and reveal a slight overall enhancement of diffraction efficiency, as well as an oscillatory dependence on the thickness of the grating.

An assumption implicit in the preceding treatments is that the structure of the holographically recorded grating is perfectly uniform throughout the depth of the recording medium. However, with any recording material having appreciable thickness and a finite absorption, the light waves used to produce the grating are attenuated inside the material, so that this assumption is no longer valid.

Diffraction by a grating in which the modulation term decreases exponentially with depth has been discussed by Kermisch [1969, 1971], Uchida [1973], Kowarschik [1976], Lederer & Langbein [1977] and Killat

[1977*a*]. For incidence at the Bragg angle the attenuation acts as a factor decreasing the effective thickness of the storage medium and reducing the maximum attainable diffraction efficiency. Away from the Bragg angle, there is an increase in the values of the secondary minima as well as a shift of the secondary maxima and minima.

Detailed experiments to study the effects of absorption in the recording medium using photographic emulsions as well as layers of dichromated gelatin have been described by Kubota [1978].

In phase holograms, it is also necessary to take into account the fact that there can be a change in the average dielectric constant. The effect of this has been analysed by Jordan & Solymar [1978], Solymar [1978] and Owen & Solymar [1980], on the assumption that the permittivity of the recording material changes during the recording process by an amount proportional to the square of the electric field. They have shown that, in general, the Bragg condition is no longer satisfied; this results in a loss in diffraction efficiency. This loss is larger for higher beam ratios, and much greater for reflection holograms than for transmission holograms.

4.8.   Criteria for thin holograms and volume holograms

As we have seen earlier, it is convenient to classify holograms into two broad categories, thin holograms and volume holograms, which correspond to the Raman-Nath and Bragg diffraction regimes respectively.

The distinction between these two regimes is commonly made on the basis of a parameter Q [Klein & Cook, 1967] which is defined by the relation

$$Q = 2\pi\lambda_0 d/n_0\Lambda^2. \tag{4.53}$$

Small values of Q ($Q < 1$) correspond to thin gratings, while large values of Q ($Q > 1$) correspond to volume gratings. However, this criterion is not always adequate.

Thus, experimental observations [Alferness, 1976; Magnusson & Gaylord, 1977] have shown that several diffracted orders can be obtained from holographic gratings recorded in lithium niobate crystals and dichromated gelatin layers under some conditions, even with relatively large values of Q.

The criteria for distinguishing between thin holograms and volume holograms have been studied in detail by several authors [Magnusson & Gaylord, 1977, 1978*a*; Moharam & Young, 1978; Benlarbi, Cooke & Solymar, 1980]. They have shown that for small levels of modulation only a single wave is diffracted, even if Q is small, so that either theory can be used, while for a large modulation higher-order waves cannot be neglected, even when Q is quite large. Consequently, as the modulation amplitude increases, an intermediate regime appears and widens. While the corresponding simple formulas can still be used in the 'thin' and 'thick' regimes, the transition boundaries between the various regimes depend on the modulation amplitude as well as the value of Q. It is possible to allow for this by the introduction of another parameter P [Nath, 1938], defined as

$$P = \lambda_0^2/\Lambda^2 n_0 n_1. \tag{4.54}$$

It can be shown that the relative power diffracted into higher-order modes is of the order of P^{-2}. Accordingly, the formulas applicable to the 'thick' regime can be used with confidence only when $P \gg 1$.

On the basis of these criteria, Moharam, Gaylord & Magnusson [1980a, b] have shown that the boundaries between the three diffraction regimes can be defined, as shown in fig. 4.10, by two curves. The boundary between the Raman–Nath and intermediate diffraction regimes corresponds closely to the curve

$$Q'\Phi = 1, \tag{4.55}$$

where $Q' = Q/\cos\theta$, θ being the angle of incidence within the grating, and Φ the modulation parameter given by (4.29).

Similarly, the boundary between the intermediate and Bragg-diffraction regimes follows the curve

$$Q'/\Phi = 20, \tag{4.56}$$

Fig. 4.10. Raman–Nath, intermediate and Bragg diffraction regimes for phase holograms [Moharam, Gaylord & Magnusson, 1980b].

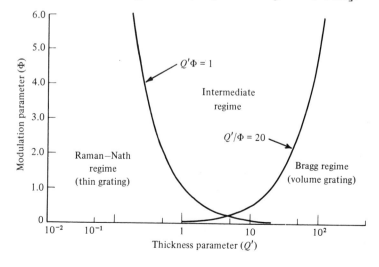

which, from (4.54), reduces to

$$P = 10. \tag{4.57}$$

4.9.   Anomalous effects with volume gratings

Some anomalous effects are also observed with volume holographic gratings which are not immediately apparent from the coupled wave theory.

Thus, Leith *et al.* [1966] observed a peak in the intensity of the transmitted wave at the same angle of illumination for which the diffracted power was also a maximum. This effect, which is analogous to the Borrmann effect in x-ray diffraction, has also been observed in gratings recorded in KBr crystals [Aristov, Shekhtman & Timofeev, 1969] and in photochromic materials [Tomlinson & Aumiller, 1975]. Forshaw [1974] found that illumination of a volume holographic grating with a narrow laser beam resulted in the emergence of four diffracted beams.

The theory of these phenomena has been worked out in detail by Russell & Solymar [1980], who have shown that for exact Bragg incidence Kogelnik's formulas can be rewritten to show the existence of two standing wave patterns parallel to the grating vector, but with a phase difference of $\pi/2$. With an amplitude grating, one of these standing waves, whose maxima coincide with the absorption maxima, is attenuated rapidly along the depth of the grating. As a result, the amplitudes of the transmitted and diffracted waves leaving the other face, which, in the limit, are only due to the other standing wave, converge to the same value.

A two-dimensional generalization of the coupled wave theory [Solymar, 1977] permits a more exact analysis for bounded beams and has been used to explain the effects observed more fully.

Other diffraction phenomena observed in very thick holograms, such as scattering rings, have been discussed by Forshaw [1975] and by Ragnarsson [1978].

4.10.   Multiply exposed holographic gratings

Due to the high angular and wavelength selectivity of volume holograms, it is possible to record several holograms in the same thick recording medium and read them out separately if their Bragg angles are sufficiently far apart. The minimum angular separation of the grating vectors for this corresponds to the condition that the maximum of the angular selectivity curve for one grating coincides with the first minimum for the other.

However, with amplitude transmission and reflection holograms, this results in a considerable loss in diffraction efficiency. On the assumption that the available dynamic range is divided equally between N gratings, it can be shown that the diffraction efficiency of each grating drops to $1/N^2$ of that for a single grating recorded in the same medium using the entire dynamic range [Collier, Burckhardt & Lin, 1971].

With volume phase gratings the situation is more complicated. Diffraction at two or more phase gratings recorded in a thick material by multiple exposures has been analysed using a thin-grating decomposition method [Alferness & Case, 1975], as well as an approach based on the coupled-wave theory [Case, 1975; Kowarschik, 1978a, b]. Their results show that with two gratings whose Bragg angles are far enough apart for coupling between the gratings to be neglected, and with linear recording, each of the gratings diffracts as if the other one did not exist. However, for higher exposures, where the recording medium is nearing saturation, reduced modulation can affect the diffraction efficiency of the individual gratings.

Coupling effects between the gratings are maximized if a common reference wave is used for all the exposures. It is then possible by a suitable choice of modulation and inclination factors to transfer the entire energy of the reconstructing wave into two or more diffracted beams and to vary their intensity ratio over a wide range. Similarly, energy can be cross coupled from one diffracted wave into the other. This makes it possible to build up beam splitters and beam couplers with specific properties.

Fig. 4.11. Changes in the orientation and spacing of the fringe planes in a hologram due to a change in the thickness of the photographic emulsion.

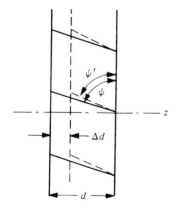

4.11.   Imaging properties of volume holograms

In the case of volume holograms, which diffract strongly at the Bragg angle, the amplitude of the reconstructed wavefront is affected by any changes of wavelength or geometry between recording and reconstruction. In addition, changes in the thickness of a photographic emulsion due to processing can result in a rotation of the fringe planes as well as a change in their spacing [Vilkomerson & Bostwick, 1967] (see fig. 4.11).

With gratings recorded with plane wavefronts it is possible to compensate for these by changing either the angle of illumination or the wavelength [Belvaux, 1975]. However, complete compensation is not possible with a hologram of a point at a finite distance, or an extended object, and this results in variations of amplitude across the reconstructed wavefront and reduced diffraction efficiency. The effects of emulsion shrinkage in volume holograms have been considered by Latta [1971c] and by Forshaw [1973], who have shown that it is possible, in this case also, to define a pupil function; this differs from the pupil function for a thin hologram defined in section 3.4, in that it involves spatial modulation of the amplitude as well as the phase.

5

Optical systems and light sources

A typical optical system for recording transmission holograms of a diffusely reflecting object is shown in fig. 5.1, while one for recording a reflection hologram is shown in fig. 5.2.

A simpler arrangement for making reflection holograms is shown in fig. 5.3. This is essentially the same as that described originally by Denisyuk [1965] in which, instead of using separate object and reference beams, the portion of the reference beam transmitted by the photographic plate is used to illuminate the object. It gives good results with specular reflecting objects and with a recording medium such as dichromated gelatin, which scatters very little light.

Making a hologram involves recording a two-beam interference pattern. The principal factors which must be taken into account in a practical setup to obtain good results are discussed in the next few sections.

5.1. Vibration isolation

Any change in the phase difference between the two beams during the exposure will result in a movement of the fringes and reduced modulation in the hologram [Neumann, 1968].

Fig. 5.1. Optical system for recording a transmission hologram.

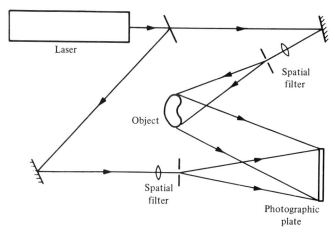

In some situations, the effects of object movement can be minimized by means of an optical system in which the reference beam is reflected from a mirror mounted on the object [Mottier, 1969]. Alternatively, if the consequent loss in resolution can be tolerated, a portion of the laser beam can be focused to a spot on the object, producing a diffuse reference beam [Waters, 1972]. Stability requirements for reflection holograms can be minimized with a setup similar to that shown in fig. 5.3 in which the surface of the object is painted with a retroreflective paint and the hologram plate is rigidly attached to it ('piggyback' holography) [Neumann & Penn, 1972].

Fig. 5.2. Typical optical arrangement for recording a reflection hologram.

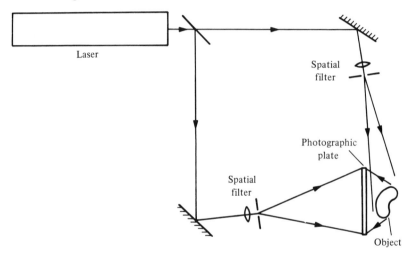

Fig. 5.3. Simple setup for making reflection holograms.

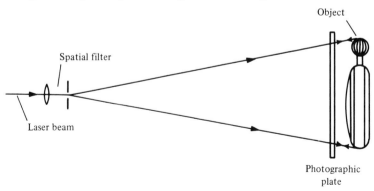

Relative motion is virtually eliminated since the plate moves with the object.

Most commonly, to avoid mechanical disturbances, all the optical components as well as the object and the recording medium are mounted on a stable surface. Acceptable results can be obtained with a concrete slab resting on inflated scooter inner tubes, but most laboratories now use a rigid optical table supported by a pneumatic suspension system, so that it has a low natural frequency of vibration (≈ 1 Hz). This has the advantage that components can be mounted on magnetic bases or bolted down to its surface.

Air currents, acoustic waves and temperature changes also cause major problems. Their effects are usually minimized by enclosing the working area.

Residual disturbances can be eliminated almost completely by a feedback system in which any motion of the interference fringes in the hologram plane is picked up by a photodetector [Neumann & Rose, 1967; MacQuigg, 1977]. Variations in its output are amplified and applied to a piezoelectric element which controls the position of a mirror in the reference-beam path to restore the path difference between the two beams to its original value.

5.2.   Fringe visibility

To produce a hologram that reconstructs a bright image, the interference pattern formed at the recording medium by the object and reference waves should have as high a contrast as possible. This is because the amplitude of the diffracted wave increases with the modulation depth of the interference pattern that is recorded.

The contrast of the interference pattern at any point in the hologram plane is measured by the fringe visibility (see Appendix A3.1) which is given by the relation

$$\mathscr{V} = (I_{max} - I_{min})/(I_{max} + I_{min}), \tag{5.1}$$

where I_{max} and I_{min} are the local maximum and minimum values of the intensity.

5.3.   Beam polarization

Most gas lasers used for holography have Brewster angle windows on the plasma tube so that the output beam is linearly polarized. Maximum visibility of the fringes will then be obtained if the angle ψ between the electric vectors in the two interfering beams is zero (see Appendix A3.1).

This condition is automatically satisfied, irrespective of the angle between the two beams, if they are both polarized with the electric vector normal to the plane of the optical table. On the other hand, if they are polarized with the electric vector parallel to the surface of the table, the angle ψ between the electric vectors is equal to the angle θ between the two beams, and, in the extreme case where the two beams intersect at right angles, the visibility of the fringes drops to zero.

With a diffusely reflecting or metallic object it is also necessary to take into account the polarization changes in the light scattered by it [Rogers, 1966; Ghandeharian & Boerner, 1978] which can result in a significant decrease in the visibility of the fringes. This can be minimized either by rotating the polarization of the reference beam so that the cross-polarized light can also interfere with it, or by using a sheet polarizer in front of the hologram plate to eliminate the cross-polarized component. Another alternative is to use a circularly polarized reference beam [Vanin, 1979].

5.4.   Beam splitters

If we assume that the interfering beams are polarized with the electric vector perpendicular to the plane of incidence, the fringe visibility is given by the relation

$$\mathscr{V} = 2|\gamma_{12}(\tau)|or/(o^2 + r^2), \tag{5.2}$$

where r and o are the amplitudes of the reference and object beams, and $\gamma_{12}(\tau)$ is the degree of coherence between them. If the beam ratio, $R = (r/o)^2$, is defined as the ratio of the irradiances of the reference and object beams, (5.2) can be rewritten as

$$\mathscr{V} = 2|\gamma_{12}(\tau)|R^{1/2}/(1 + R). \tag{5.3}$$

The fringe visibility is obviously a maximum when $R = 1$. However, the wave scattered by a diffusely reflecting object exhibits quite strong variations in amplitude (see Appendix A4.1). Hence, in hologram recording it is usually necessary to work with a value of $R \gg 1$ to avoid nonlinear effects.

To optimize the visibility of the fringes at the hologram plane it should be possible to vary the ratio of the power in the beam illuminating the object to that in the reference beam.

A convenient way to do this is to use a beam splitter consisting of a disc coated with a thin aluminium film whose reflectivity is a linear function of the azimuth. Such a beam splitter must be used in the unexpanded laser beam because of the gradient of reflectivity across a larger beam. Since an aluminium film typically absorbs about 30 per cent of the energy incident on it, this limits its use to moderate laser powers (< 500 mW).

A much better variable-ratio beam splitter, with a uniform field and very low insertion loss, uses a polarizing prism to divide the incident beam into two orthogonally polarized components [Caulfield & Beyen, 1967]. A typical setup for this is shown in fig. 5.4. The ratio of the transmitted and reflected powers is given by the relation

$$I_{\text{trans}}/I_{\text{refl}} = \cot^2 \psi, \tag{5.4}$$

where ψ is the angle which the incident electric vector makes with the vertical. This ratio can be conveniently controlled by using a half-wave plate to rotate the direction of polarization of the input beam. Another fixed half-wave plate in the transmitted beam is used to bring the electric vector of this beam back to vertical.

5.5.   Beam expansion

The laser beam has to be expanded to illuminate the object and the plate on which the hologram is recorded. Usually this is done with microscope objectives.

If the laser is oscillating in the TEM_{00} mode, the beam has a Gaussian profile so that the amplitude at a point at a radial distance r from the centre

Fig. 5.4. Variable-ratio beam splitter using a polarizing prism.

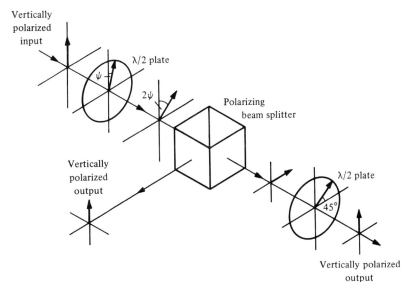

of the beam is

$$a(r) = a(0) \exp(-r^2/w^2), \tag{5.5}$$

where $a(0)$ is the amplitude at the centre of the beam and w is the beam radius (the radius at which the amplitude drops to $1/e$ of its maximum value).

Accordingly, $P(r)$ the laser power within a circle of radius r is given by the relation

$$P(r) = \int_0^r I(r)2\pi r \, dr, \tag{5.6}$$

where $I(r) = |a(r)|^2$ is the intensity at a radial distance r from the centre of the beam; this can be written as

$$P(r) = P_{tot}[1 - \exp(-2r^2/w^2)], \tag{5.7}$$

where P_{tot} is the total power in the beam.

From (5.5) and (5.7) it follows that

$$P(r)/P_{tot} = 1 - [I(r)/I(0)]. \tag{5.8}$$

This is a useful relation giving the loss in power that must be tolerated for a given degree of uniformity of illumination.

Due to the high coherence of laser light, the expanded beam invariably exhibits diffraction patterns (spatial noise) due to scattering from dust particles on the optical surfaces in the beam path. To eliminate these, a pinhole is placed at the focus of the microscope objective. If this pinhole has a radius ρ, spatial frequencies higher than $\xi = \rho/\lambda f$, where λ is the wavelength of the light and f is the focal length of the microscope objective, are blocked. These higher spatial frequencies mainly represent noise, so that the transmitted beam has a smooth Gaussian profile.

A proper choice of the size of the pinhole can ensure that the power loss is minimal. Thus, it follows that the amplitude in the focal plane of the microscope objective, obtained from the two-dimensional Fourier transform of (5.5), is

$$A(\rho) = A(0) \exp(-\pi^2 w^2 \rho^2/\lambda^2 f^2), \tag{5.9}$$

where ρ is the radial distance from the centre of the beam.

Typically, with an argon-ion (Ar^+) laser, $\lambda = 514$ nm and $w = 0.8$ mm, and, with a 16 mm microscope objective and a 10 μm diameter pinhole, the amplitude at the edge of the pinhole is only 0.08 of that at the centre. Hence, from (5.8), more than 99 per cent of the total power in the beam is transmitted through the pinhole.

5.6.   Exposure control

An accurate spot photometer is required to measure the irradiances due to the object and reference beams in the hologram plane, so as to set the beam ratio R at a suitable value. Because of the limited dynamic range of photographic materials used for holography, the object illumination should be adjusted so that the irradiance in the hologram plane due to it is reasonably uniform. In addition, precise control of the exposure is required to ensure good diffraction efficiency and avoid nonlinear effects. It is not enough to maintain a specified exposure time because the laser output can fluctuate during the exposure. To overcome this problem it is convenient to use an electronic exposure-control unit which integrates the irradiance in the hologram recording plane and closes the shutter at a preset value of radiant exposure [Lin & Beauchamp, 1970*a*]. A more sophisticated system which is suitable for multicolour holography has been described by Oreb & Hariharan [1981].

5.7.   Coherence requirements

In order to obtain maximum fringe visibility, it is also essential to use coherent illumination. Gas lasers provide an intense source of highly coherent light and are therefore used almost universally in optical holography.

Spatial coherence is automatically ensured if the laser oscillates in a single transverse mode, preferably the lowest order or TEM_{00} mode, since this is inherently the most stable and gives most uniform illumination over the field. Normally, this is no problem since most gas lasers are designed to operate in this mode [Bloom, 1968]. However, they are not usually designed for single-frequency operation, which would imply that they should also oscillate in only one longitudinal mode. The temporal coherence of the light from most lasers is therefore limited by their longitudinal mode structure.

To obtain a satisfactory hologram, the maximum optical path difference between the object and reference beams in the recording setup must be less than the coherence length (see Appendix A3.4) of the light from the laser. With an extended object, the holodiagram [Abramson, 1969] (see section 14.9) can be used to optimize the layout.

5.8.   Temporal coherence of laser light

The simplest form of resonant cavity for a laser is made up of a pair of mirrors separated by a distance L, though in some cases, where operation on more than one line is possible, a wavelength selector prism may also be necessary as shown in fig. 5.5. The resonant frequencies of such a cavity are given by the expression

$$\nu_n = n(c/2L),\qquad(5.10)$$

where n is an integer and c is the speed of light. However, as shown in fig. 5.6, the laser can oscillate only at those frequencies within the gain curve of the active medium at which the gain is adequate to overcome the cavity losses.

The width of the individual modes depends on the losses as well as the mechanical stability of the cavity structure and is typically about 3 MHz. Accordingly, if a laser is made to oscillate in a single longitudinal mode, the coherence length would be of the order of 100 metres.

Since, in frequency space, the width of the individual modes is much less

Fig. 5.5. Optical system of a typical Ar$^+$ laser.

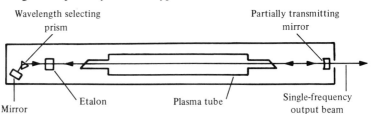

Fig. 5.6. Oscillation frequencies of a laser with (a) a long resonant cavity and (b) a short resonant cavity.

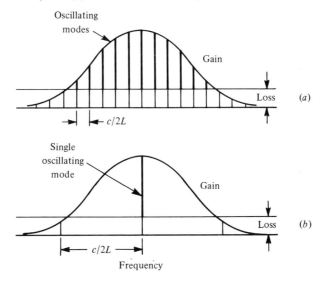

than their separation (which may range from 600 MHz for a 25 cm cavity down to 75 MHz for a 2 m cavity), the power spectrum of a laser oscillating in N longitudinal modes can be represented by N equally spaced delta functions. If we assume that the total power is equally divided between the N modes, the power spectrum can be written as

$$S(v) = \sum_{n}^{n+N-1} \delta(v - v_n). \tag{5.11}$$

The degree of temporal coherence can then be evaluated from (A3.14), and, for a path difference p, is

$$\gamma_{12}(p) = \left| \frac{\sin(N\pi p/2L)}{N \sin(\pi p/2L)} \right|. \tag{5.12}$$

This is a periodic function whose first zero occurs when

$$p = 2L/N, \tag{5.13}$$

corresponding to the effective coherence length. Fringes with acceptable visibility can usually be obtained for path differences that are less than half the coherence length.

It is apparent from (5.13) that the existence of more than one longitudinal mode in a laser reduces the coherence length severely. A short coherence length is troublesome, since it makes it essential to equalize the mean optical paths of the object and reference beams and restricts the maximum depth of the object field that can be recorded.

The simplest way to force a laser to operate in a single longitudinal mode is to use a very short cavity so that the spacing of the longitudinal modes is greater than the width of the gain profile over which oscillation is possible. This occurs when

$$c/2L \geqslant \Delta v, \tag{5.14}$$

where Δv is the width of the gain profile (typically 1.7 GHz for a helium–neon (He–Ne) laser, 3.5 GHz for an Ar^+ laser). However, the power available from such a short cavity is extremely limited.

The most common method of ensuring single frequency operation is to use an intracavity etalon as shown in fig. 5.5. In effect the laser is now made up of two resonant cavities, and only those modes which are common to both cavities, as shown in fig. 5.7, have low enough losses for oscillation to be possible. If the length of the etalon is made short enough that it satisfies (5.14), it can support only one mode, and single-frequency operation is obtained.

If the etalon is to act as a simple transmission filter, it must be tilted to decouple it from the laser cavity. It can then be tuned to maximize the output power by tilting it further so that its resonant frequency corresponds to the peak of the gain curve [Hercher, 1969]. An alternative method of tuning which gives better frequency stability and efficiency is to mount the etalon within an oven which can be maintained at any desired temperature. Decoupling is easier if an etalon with concentric surfaces is used [Hariharan, 1982].

5.9.   Gas lasers

The light source most commonly used for holography is a gas laser. Gas lasers, in general, are cheaper, easier to operate and have better coherence characteristics than other types of lasers. The range of useful wavelengths and typical output powers available at these wavelengths with the most commonly used gas lasers are summarized in Table 5.1.

When working with any laser, adequate safety measures must be taken to avoid eye damage (see ANSI, 1980; BSI, 1983). Even with relatively low-power gas lasers (≈ 1 mW) the beam should not enter the eye directly. This is because the beam is focused into a very small spot on the retina resulting in a power density about 10^5 times that at the cornea. With medium power lasers (< 100 mW) care must also be taken to avoid stray reflections, but the

Fig. 5.7. Normal multifrequency output, etalon transmittance and single-frequency output with an intracavity etalon for a typical Ar$^+$ laser.

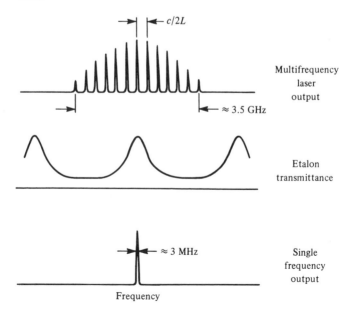

$c/2L$

Multifrequency laser output

≈ 3.5 GHz

Etalon transmittance

≈ 3 MHz

Single frequency output

Frequency

risk of eye damage is minimal once the beam has been diffused or expanded. However, at power levels > 1 W, viewing even a light-coloured diffusing surface illuminated by the unexpanded beam is dangerous, since the surface intensity is comparable to that of the sun. With such lasers, as well as with pulsed lasers (see section 5.10) safety glasses must be worn.

For a simple holographic set up, the He–Ne laser is by far the most economical choice. It operates on a single line at 633 nm, does not require water cooling and has a long life. However, depending on their power, commercial He–Ne lasers oscillate in three to five longitudinal modes, and the coherence length is limited.

In contrast to the He–Ne laser, the Ar$^+$ laser essentially has a multiline output but can be made to operate on a single line by replacing the reflecting end mirror by a prism and mirror assembly. It is also relatively easy to obtain single-frequency operation with an etalon. Argon–ion lasers can give high power output and an extended coherence length in the blue or green regions of the spectrum, the two strongest lines being at 488 nm and 514 nm.

The krypton-ion (Kr$^+$) laser is very similar in its construction and characteristics to the Ar$^+$ laser and is a useful replacement for the He–Ne laser, where high output power and an extended coherence length are required at the red end of the spectrum (647 nm).

The helium–cadmium (He–Cd) laser provides a stable source at a relatively short wavelength (442 nm). It is very useful with recording materials such as photoresists (see section 7.3), whose increased sensitivity at this wavelength makes up for the lower power available.

Table 5.1. *Output wavelength and output power of gas lasers*

Wavelength nm	Laser	Typical power mW	Colour
442	He–Cd	25	Violet
458	Ar$^+$	200	Blue–violet
476	Kr$^+$	50	Blue
477	Ar$^+$	400	Blue
488	Ar$^+$	1000	Green–blue
514	Ar$^+$	1400	Green
521	Kr$^+$	70	Green
633	He–Ne	2–50	Red
647	Kr$^+$	500	Red

5.10.   Pulsed lasers

With a pulsed laser, problems connected with the stability of the recording system are virtually eliminated [Jacobson & McClung, 1965; Siebert, 1967]. The ruby laser is still the most widely used type of pulsed laser for optical holography [Koechner, 1979a, b] mainly because of the fairly large output energy available (up to 10 J per pulse) and the wavelength of the light emitted (694 nm) which is fairly well matched to the peak sensitivity of photographic materials available for holography.

While neodymium-doped yttrium aluminium garnet (Nd:YAG) lasers can operate more efficiently and at a higher pulse repetition rate, and have also been used for holography [Gates, Hall & Ross, 1970; Bates, 1973], they have the disadvantages that the energy per pulse is usually less, and a frequency doubling crystal must be used to convert the infrared output to visible light (530 nm). However, an Nd:YAG laser system for pulsed holography of large scenes has been described by Andreev, Vorzobova, Kalintsev & Staselko [1980].

The active medium in a ruby laser is a rod, typically 5–10 mm in diameter and 75–100 mm in length, made of synthetic sapphire (Al_2O_3) doped with 0.05 per cent of Cr_2O_3. When this is mounted in a suitable optical resonator, as shown in fig. 5.8, and pumped with a xenon flashlamp, it emits a series of pulses lasting approximately 250 μs, and polarized with the electric vector normal to the axis of the rod and the crystal c-axis [Lengyel, 1971].

For holography it is necessary to use a Q-switch in the cavity. This is a fast-acting optical shutter, which is normally closed so that the laser cannot oscillate when the flash lamp is fired and a large population is built up in the upper level. When the switch opens, the stored energy is released in a very short pulse with very high peak power.

Fig. 5.8. Optical system of a Q-switched ruby laser for holography.

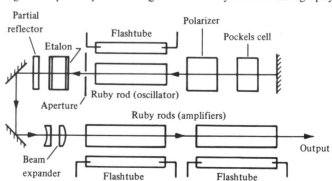

The most useful type of Q-switch for holography is the Pockels cell, since it permits precise control of the timing and number of output pulses. This uses a crystal which exhibits birefringence when an electric field is applied to it. The Pockels cell is located in the laser cavity as shown in fig. 5.8, with its principal axes at $45°$ to the direction of polarization of the laser beam, between the end reflector and a polarizer which is oriented to pass the laser beam. When the flashlamp is fired, a voltage is applied to the Pockels cell, producing a phase shift of $\pi/2$ between the two transmitted components. Any light from the laser rod transmitted by the Pockels cell and reflected back by the end mirror then emerges polarized at $90°$ to its original direction and is blocked by the polarizer. When the voltage on the Pockels cell is switched off, towards the end of the flashlamp pulse, the combination of the Pockels cell and the polarizer transmits freely, allowing laser oscillations to build up, so that a very short pulse, typically with a duration of 15 ns, is produced.

Because of the high gain of a ruby rod in a typical optical resonator, such a laser normally oscillates in a large number of transverse modes, and the spatial coherence of the output is very poor. To ensure operation only in the TEM_{00} mode, an aperture with a diameter of about 2 mm is inserted at a suitable point in the resonator, as shown in fig. 5.8.

Again, because of the relatively large width of the fluorescence line (≈ 400 GHz), a ruby laser normally oscillates in as many as 100–200 longitudinal modes and has a very short coherence length (< 1 mm). To obtain an increased coherence length, it is necessary to use a Fabry–Perot etalon in the cavity. This is normally a plate of fused silica, or sapphire, about 2 or 3 mm thick. True single-mode operation can be obtained with resonant reflectors consisting of two plates separated by a suitable distance, along with the use of a Q-switch with a relatively slow rise time, which prevents the modes with lower gain from building up to any appreciable extent before laser action occurs [McClung, Jacobson & Close, 1970; Young & Hicks, 1974].

Since the single-mode output from a ruby oscillator is limited (≈ 0.04 J), high-power ruby lasers use one or more additional ruby rods as amplifiers to boost the output while preserving spatial and temporal coherence. For maximum gain with stable operation, the amplifier rod is set so that its c-axis is parallel to that of the oscillator and may be optically isolated from it by a spatial filter consisting of a sapphire pinhole placed at the common focus of two lenses. The focal lengths of these lenses are chosen so that the beam from the oscillator is expanded to fill the amplifier rod, which may typically have a diameter of 20 mm and a length of 200 mm. An output of

1 J is possible with a single amplifier stage, and 10 J with two stages of amplification, with a coherence length of 5–10 metres.

5.11. Holography with pulsed lasers

Because of the very short duration of the output pulse from a Q-switched ruby laser (about 15 ns), it is possible to record a hologram of a moving object.

The drop in the visibility of the hologram fringes due to the movement of the object can be evaluated either from the change in the optical path or from the shift in optical frequency due to the Doppler effect [Mallick, 1975]. The permissible object displacement depends very much on the geometry of the recording system, but, in general, object velocities up to a few metres per second can be tolerated. The holodiagram [Abramson, 1969] (see section 14.9) can be used conveniently to minimize the sensitivity of the system to movement of the object in a particular direction.

Because of the high power density ($> 10^{12}$ W/m^2) in the undiverged laser beam, it is essential to use a beam splitter with low insertion loss and multilayer dielectric mirrors instead of metallized mirrors, which have a damage threshold of only about 10^{10} W/m^2. It is not possible to use microscope objectives with pinhole spatial filters to expand the beams; instead, simple negative lenses are used. All optical surfaces must be kept clean and free of dust particles, which can lower the damage threshold appreciably. Adequate safety precautions, including the use of protective eyewear, are essential [see ANSI, 1980; BSI, 1983].

Fig. 5.9. Typical setup for recording a hologram of a human subject using a Q-switched ruby laser [Ansley, 1970].

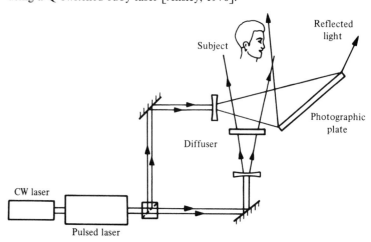

The Q-switched ruby laser also makes it possible to record a hologram of a living human subject [Siebert, 1968]. To avoid eye damage, it is important to see that the energy density at the retina of the subject does not exceed the safety limit. This is possible if, instead of illuminating the subject directly with the expanded laser beam, it is allowed to fall on a fairly large diffuser which scatters the light effectively constituting an extended source. A typical optical arrangement is shown in Fig. 5.9 [Ansley, 1970]. The dimensions of the diffuser should be chosen so that the integrated radiance of the diffuser, measured at the cornea of the subject, is less than the maximum permissible exposure for an extended source of 700 J/m²/sr; the undeviated energy density should also be less than 5 mJ/m²/ [see ANSI, 1980; BSI, 1983]. In addition, it is necessary to ensure that the specular reflection of the reference beam from the surface of the photographic plate is directed away from the subject.

Closer control of the illuminated area with less wastage of light is possible if a holographic scatter-plate is used instead of a diffuser [Webster, Tozer & Davis, 1979]. However, such scatter plates are not completely effective as diffusers and it is essential, if they are used for portraiture, that the setup should be modified so that the directly transmitted beam does not fall on the subject.

5.12.   Dye lasers

Even though dye lasers have not been used widely for holography, they appear in many ways to be the fulfilment of an experimenter's dream. Early dye lasers were built for pulsed operation, but it was not long before it was shown that continuous-wave operation could be obtained with a thin flowing layer of a dye by focusing the beam from a high-power Ar^+ laser on it.

Because of the very large number of dyes available, it is possible to change the operating wavelength of a dye laser merely by switching dyes. In addition, with a given dye, it is possible to tune the output over a range of wavelengths (usually about 50–80 nm) by incorporating a wavelength selector, such as a diffraction grating, or a birefringent filter, in the laser cavity. Operation in a single longitudinal mode can also be obtained by means of an intracavity etalon. Dye-laser systems are now available which can give a single-frequency output of a few hundred milliwatts at any desired wavelength within the visible spectrum.

6

The recording medium

Before we look at the different recording materials which have been used for optical holography, it is necessary to define the most important characteristics of a holographic recording medium.

6.1. Macroscopic characteristics

Any material used to record a hologram must respond to exposure to light (after additional processing, where necessary) with a change in its optical properties. The complex amplitude transmittance of such a material can be written in the most general manner as

$$\mathbf{t} = \exp(-\alpha d)\exp[-\mathrm{i}(2\pi nd/\lambda)],$$
$$= |\mathbf{t}|\exp(-\mathrm{i}\phi), \tag{6.1}$$

where α is the absorption constant of the material, d is its thickness and n is its refractive index. Accordingly, as mentioned earlier, holographic recording materials can be classified, for convenience, either as pure amplitude-modulating materials if only α changes with the exposure, or as pure phase-modulating materials if $\alpha \approx 0$ and either n or d changes with the exposure.

The response of the recording material, defined by (6.1), can be described in these two limiting cases by curves of amplitude transmittance *versus* exposure ($|\mathbf{t}| - E$ curves) as shown in fig. 6.1, or curves of the effective phase shift against exposure ($\Delta\phi - E$ curves) as shown in fig. 6.2.

6.2. The modulation transfer function

While curves such as those in figs. 6.1 and 6.2 describe the behaviour of a recording material quite satisfactorily on a macroscopic scale, they are not, by themselves, adequate to predict the response of the material on a microscopic scale.

As discussed in section 5.2 the intensity at any point on the hologram recording medium is given by a relation of the form

$$I = \langle I\rangle[1 + \mathscr{V}\cos(\phi_r - \phi_o)], \tag{6.2}$$

where $\langle I\rangle$ is the average intensity, $\mathscr{V}$ is the visibility of the interference fringes formed by the object and reference beams, and ϕ_o and ϕ_r are their phases at this point.

The actual intensity distribution to which the material is exposed is different, because there is always a certain amount of lateral spreading of light within the recording medium, determined by its scattering properties and its coefficient of absorption. Because of this, the actual modulation of the intensity within the recording material is always less than that in the original interference pattern.

For any spatial frequency s, the ratio of $\mathscr{V}'(s)$, the actual modulation of

Fig. 6.1. Curve of amplitude transmittance $|t|$ against exposure (E) for a recording material.

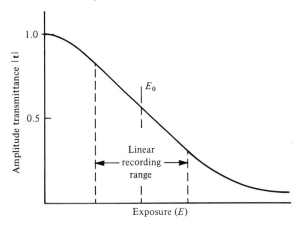

Fig. 6.2. Curve of phase shift $(\Delta\phi)$ against exposure (E) for a recording material.

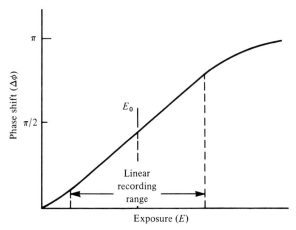

the intensity distribution within the material, to $\mathscr{V}(s)$, the input modulation, is termed the modulation transfer function (or MTF) $M(s)$, so that

$$M(s) = \mathscr{V}'(s)/\mathscr{V}(s). \tag{6.3}$$

This parameter, which is normally less than unity, serves to characterize the relative response of the material at different spatial frequencies.

6.3.   Diffraction efficiency

The diffraction efficiency ε of a hologram has been defined (see section 3.6) as the ratio of the power diffracted into the desired image to that illuminating the hologram.

For an exposure time T, the effective exposure of the recording material is, from (6.2) and (6.3),

$$\begin{aligned}
E &= TI, \\
&= T\langle I \rangle [1 + \mathscr{V}(s)M(s)\cos(\phi_r - \phi_o)], \\
&= \langle E \rangle [1 + \mathscr{V}(s)M(s)\cos(\phi_r - \phi_o)]. \tag{6.4}
\end{aligned}$$

If we assume linear recording, as in (2.2), the amplitude transmittance of the recording medium can be written as

$$\mathbf{t} = \langle \mathbf{t} \rangle + \beta \mathrm{d}E, \tag{6.5}$$

where $\langle \mathbf{t} \rangle$ is the amplitude transmittance for the average exposure $\langle E \rangle$ and β is the value of $(\mathrm{d}t/\mathrm{d}E)$ at $\langle E \rangle$. Accordingly, the amplitude transmittance of the hologram is

$$\mathbf{t} = \langle \mathbf{t} \rangle + \beta \langle E \rangle \mathscr{V}(s)M(s)\cos(\phi_r - \phi_o). \tag{6.6}$$

When the hologram is illuminated once again with the original reference wave r, the amplitude of the reconstructed image is

$$u_3 = (1/2)r\beta \langle E \rangle \mathscr{V}(s)M(s), \tag{6.7}$$

and its intensity is

$$I_3 = (1/4)|r|^2 [\beta \langle E \rangle \mathscr{V}(s)M(s)]^2. \tag{6.8}$$

Accordingly, the diffraction efficiency of the hologram is

$$\varepsilon = (1/4)[\beta \langle E \rangle \mathscr{V}(s)M(s)]^2. \tag{6.9}$$

Now, since $\beta = (\mathrm{d}t/\mathrm{d}E)$,

$$\begin{aligned}
\beta E &= E(\mathrm{d}t/\mathrm{d}E), \\
&= \log_{10}e \cdot (\mathrm{d}t/\mathrm{d}\log E), \\
&= 0.434\Gamma, \tag{6.10}
\end{aligned}$$

where $\Gamma = (\mathrm{d}t/\mathrm{d}\log E)$.

The diffraction efficiency is proportional to the square of Γ, the gradient of the **t** *versus* log *E* curve, as well as to the squares of $\mathscr{V}(s)$, the input modulation, and *M(s)*, the MTF [Biedermann, 1969]. The maximum diffraction efficiency is obtained where the slope of the **t** *versus* log *E* curve is steepest. This is usually at a slightly higher exposure than that corresponding to the steepest part of the **t** *versus E* curve.

A method of characterizing recording materials for holography based on (6.9) has also been described by Lin [1971]. For an ideal recording material, plots of $\sqrt{\varepsilon}$ *versus* $\langle E \rangle$ with $\mathscr{V}(s)$ as a parameter and plots of $\sqrt{\varepsilon}$ *versus* $\mathscr{V}(s)$ with $\langle E \rangle$ as a parameter should be straight lines. The departure of the characteristics of any real material from the ideal can therefore be easily seen by comparing the actual measured curves with these. Apart from the maximum diffraction efficiency, or the exposure needed to obtain a given diffraction efficiency, these curves also make it possible to determine the range of fringe visibility $\mathscr{V}(s)$ or beam ratio *R* for which the hologram recording is linear (indicated by the straight line portion of the $\sqrt{\varepsilon}$ *versus* $\mathscr{V}(s)$ curve at constant $\langle E \rangle$) and the value of average exposure representing the best compromise between linearity and efficiency.

6.4.   Image resolution

With an ideal recording material, the resolution of the image is determined only by the dimensions of the hologram. However, with any practical recording material, its MTF will, in general, affect the resolution as well as the intensity of the reconstructed image [Kozma & Zelenka, 1970].

This is because, in most holographic systems, the spatial frequency *s* varies over the hologram. It then follows from (6.9) that, due to the corresponding variations in the MTF of the recording medium, the diffraction efficiency of the hologram will vary over its aperture. Wherever this variation is appreciable, it must be taken into account in evaluating the imaging properties of the hologram.

The only exceptions to this situation are when the image is formed at or near the hologram or with a Fourier hologram (see section 2.3). In the latter case, the spatial frequency *s* is constant over the entire hologram. As a result, the MTF of the recording medium only affects the intensity of the image; its resolution is determined by the aperture of the hologram.

At the other extreme, where the object and the reference source are at different distances from the hologram, and the hologram is quite large, it is possible for the spatial frequency *s* to exceed the resolution limit of the recording material over part of the hologram. This limits the useful aperture of the hologram and, hence, the resolution of the image.

A simple method to visualize the effects of the MTF of the recording material, for a given point on the hologram, makes use of the concept of a fictitious mask located in the object beam [van Ligten, 1966]. The amplitude transmittance of this mask is proportional to the MTF, ranging from unity at a point on the object in line with the reference source (and hence corresponding to zero spatial frequency at the hologram) down to complete opacity at points at a large enough lateral separation from the reference source. More exactly, the effects of the MTF of the recording material can be taken into account, as was done for the effects of angular and wavelength selectivity in a thick hologram (see section 4.11), by defining a generalized space-variant pupil function, involving, in this case, the local spatial frequency and the MTF of the recording material [Kozma & Zelenka, 1970; Jansson, 1974].

6.5.   Nonlinearity

Consider the interference of an object wave o and a reference wave r to produce an amplitude hologram. Assuming linear recording as in (2.2), the amplitude transmittance of the hologram is

$$\mathbf{t} = \mathbf{t}_0 + \beta T[rr^* + oo^* + r^*o + ro^*]. \tag{6.11}$$

However, in practice, the assumption of linear recording is not valid when the fluctuating terms on the right hand side of (6.11) are comparable with the bias term.

This is usually the situation with phase holograms because of their intrinsic nonlinearity. Even if the phase shift produced by the recording medium is strictly proportional to the irradiance in the interference pattern, the complex amplitude transmittance at any point in a phase hologram is

$$\begin{aligned}\mathbf{t}(x, y) &= \exp(-i\phi), \\ &= 1 - i\phi - (1/2)\phi^2 + (1/6)i\phi^3 \dots.\end{aligned} \tag{6.12}$$

If the phase modulation is increased to obtain high diffraction efficiency, the effects of the higher-order terms in (6.12) cannot be neglected.

The effects of nonlinear recording were first investigated by Kozma [1966] and Friesem & Zelenka [1967] for simple objects, and by Goodman & Knight [1968] for a diffusely reflecting object, using the characteristic function method of communication theory. A simpler approach [Bryngdahl & Lohmann, 1968c] is to assume that the amplitude transmittance of the recording material can be represented by a polynomial. This can be written as

$$\mathbf{t} = \mathbf{t}_0 + \beta_1 E + \beta_2 E^2 + \ldots,$$
$$= \mathbf{t}_0 + \beta_1 T[rr^* + oo^* + r^*o + ro^*]$$
$$+ \beta_2 T[rr^* + oo^* + r^*o + ro^*]^2 + \ldots, \qquad (6.13)$$

if higher-order terms are neglected. When the hologram is illuminated again by the reference wave r, which, for simplicity, can be assumed to be a plane wave of unit amplitude, the complex amplitude of the wave transmitted by the hologram is

$$u = \text{linear terms}$$
$$+ \beta_2 T^2[(oo^*)^2 + o^2 + o^{*2} + 2o^2 \, o^* + 2oo^{*2}]. \qquad (6.14)$$

The immediate result of nonlinear recording is, therefore, the production of additional spurious terms.

A comparison of (6.14) and (6.11) shows that the term involving $(oo^*)^2$ in (6.14), like the term involving oo^* in (6.11), corresponds to a halo surrounding the directly transmitted beam. However, the spatial frequency spectrum of this term, which is obtained by convolving the spectrum of oo^* with itself, has twice the width of the spectrum of oo^*. This corresponds to a doubling of the angular width of the halo surrounding the directly transmitted beam.

The second term o^2 is the square of the complex amplitude of the object wave at the hologram. The mean direction of the corresponding diffracted wave makes an angle with the axis which is approximately twice that made by the object wave; this diffracted wave also has a curvature twice that of the object wave. Accordingly, it gives rise to a higher-order primary image. Similarly, the third term o^{*2} can be interpreted as a higher-order conjugate image.

The fourth and fifth terms, which result in diffracted wavefronts with complex amplitudes proportional to $2o^2o^*$ and $2oo^{*2}$, can be shown to be intermodulation terms giving rise to false images. With an extended diffusely reflecting object, the effect of these false images is to create a 'noise halo' about the true image. This can be understood quite readily if we realize that the spatial-frequency spectrum of the wave corresponding to the fourth term is obtained by convolving the spatial-frequency spectrum of the object with the spatial-frequency spectrum of the speckle pattern produced by the object wave in the hologram plane.

Further studies using a power series technique have been made by Kozma [1968*b*] and by Kozma, Jull & Hill [1970]. Their results can be used to calculate the ratio of the intensity in the reconstructed image to that due to the nonlinear noise background, as well as the shape of the noise distribution for a diffusely illuminated object. Unfortunately, the expressions obtained are quite complicated. This difficulty is partly overcome in an alternative treatment by Tischer [1970] based on the use of Chebyshev polynomials. With simple objects such as lines and points, the terms of the series represent corresponding pairs of ghost images, so that, in principle, the optimum operational conditions can be evaluated readily.

However, the polynomial method has limitations, as pointed out by Velzel [1973], who was able to develop a complete theory of nonlinear holographic image formation using a modified transform method, and derive analytical expressions for the efficiency and the image contrast for recording media with an exponential characteristic (a linear phase hologram) and a binary characteristic. A generalization of this theory which can be used where the transfer function of the recording medium is complex has been outlined by Ghandeharian & Boerner [1977].

6.6.   Effect of hologram thickness

Intermodulation effects would make it almost impossible to produce bright holographic images of good quality but for the fact that, in a volume hologram, intermodulation noise is reduced significantly by the angular selectivity of the hologram (see sections 4.4 and 4.5).

Consider a simple object consisting of only two points, which gives rise to two sets of interference fringes with the reference beam having spatial frequencies s_{01} and s_{02}. The intermodulation terms then have the general form $ps_{01} \pm qs_{02}$, where p and q are positive integers. Most of the resultant spatial frequencies are therefore significantly different from those corresponding to the object points. Qualitatively, it is apparent that when the hologram is illuminated once again with the same reference beam used to record it, the Bragg condition will not be satisfied for these intermodulation frequencies [Upatnieks & Leonard, 1970].

A detailed analysis of intermodulation noise in a volume hologram due to nonlinear recording has been made by Guther & Kusch [1974], using the coupled wave theory (see section 4.3). Their results show that the noise is effectively limited by the thickness of the recording medium as well as by the aperture of the hologram. A simpler analysis by Hariharan [1979a] shows that if the angle between the two beams in the recording setup is large enough for the diffracted beams corresponding to different orders not to overlap, the signal-to-noise ratio should improve by a factor approximately equal to $(\psi/\Delta\theta)$ where $2\Delta\theta$ is the width of the passband of the angular selectivity function, and ψ is the angular extent of the object.

6.7. Noise

In optical holography, noise is a convenient term for non-image-forming light which is diffracted or scattered in the same general direction as the reconstructed image.

If we exclude noise generated by nonlinear effects, which has been discussed in section 6.5, the other source of noise with recording materials such as photographic emulsions is scattered light from the individual grains, which are distributed in the emulsion layer in a random manner [Goodman, 1967; Kozma, 1968a].

The noise spectrum of a photographic material can be studied with the optical system shown in fig. 6.3. If a uniformly exposed photographic plate with an amplitude transmittance $t(x, y)$ is placed in front of the lens and illuminated with a collimated beam of monochromatic light, the diffracted amplitude at any point (x_f, y_f) in the back focal plane of the lens is (see Appendix A2.3)

$$a(x_f, y_f) = (ia/\lambda f) \exp\left[(-i\pi/\lambda f)(x_f^2 + y_f^2)\right] \mathbf{T}(\xi, \eta), \qquad (6.15)$$

where a is the amplitude of the plane wave illuminating the plate, f is the focal length of the lens and $t(x, y) \leftrightarrow \mathbf{T}(\xi, \eta)$; ξ and η are spatial frequencies defined by the relations $\xi = x_f/\lambda f$, $\eta = y_f/\lambda f$.

The intensity at this point is then

$$\begin{aligned} I(x_f, y_f) &= |a(x_f, y_f)|^2, \\ &= (a^2/\lambda^2 f^2)|\mathbf{T}(\xi, \eta)|^2. \end{aligned} \qquad (6.16)$$

Fig. 6.3. Arrangement for measuring the noise spectrum of photographic materials used for holography.

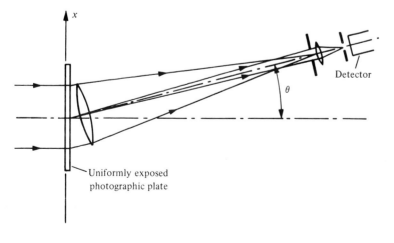

x

θ

Detector

Uniformly exposed
photographic plate

If this is averaged over a large enough area to eliminate local fluctuations due to speckle, it can be written as

$$I(\xi, \eta) = (a^2/\lambda^2 f^2)S(\xi, \eta), \qquad (6.17)$$

where $S(\xi, \eta)$ is the power spectrum of the transmittance of the plate.

To eliminate effects due to diffraction at the edges of the sample, a second lens is normally used to image the sample on to another aperture placed in front of the detector. The effective area of the sample is then the projection of this aperture on the sample. Since this is proportional to $1/\cos \theta$, it also compensates for the decrease in the effective passband of the collecting aperture [Biedermann & Johansson, 1975].

6.8.   Signal-to-noise ratio with coherent illumination

In computing the signal-to-noise ratio in the image reconstructed by a hologram, it is necessary to take into account the fact that it is the amplitudes of the signal and the noise which have to be added, since they are both encoded on the same coherent carrier [Goodman, 1967].

Consider the reconstructed image of an object consisting of a bright patch on a dark background. Let the intensity due to the nominally uniform signal be I_i, while that of the randomly varying background is I_N. The noise N in the bright area can be defined as the variance of the resulting fluctuations of the intensity, and is given by the relation

$$N = [\langle I^2 \rangle - \langle I \rangle^2]^{1/2}, \qquad (6.18)$$

where I is the intensity at any point and $\langle I \rangle$ is the average intensity.

Since a_N the complex amplitude of the background has circular statistics (see Appendix A4), the mean value of terms involving a_N or a_N^* as well as powers of a_N and a_N^*, other than those involving only $|a_N|^2$, is zero. Hence, if a_i is the complex amplitude of the signal, the average intensity is

$$\begin{aligned} \langle I \rangle &= \langle (a_i + a_N)(a_i^* + a_N^*) \rangle, \\ &= \langle I_i + I_N + a_i a_N^* + a_i^* a_N \rangle, \\ &= I_i + \langle I_N \rangle. \end{aligned} \qquad (6.19)$$

Similarly,

$$\begin{aligned} \langle I^2 \rangle &= \langle (I_i + I_N + a_i a_N^* + a_i^* a_N)^2 \rangle, \\ &= \langle I_i^2 + I_N^2 + 4 I_i I_N \rangle, \\ &= I_i^2 + \langle I_N^2 \rangle + 4 I_i \langle I_N \rangle. \end{aligned} \qquad (6.20)$$

However, from (A4.4) and (A4.5), $\langle I_N^2 \rangle = 2 \langle I_N \rangle^2$, so that

$$\langle I^2 \rangle = I_i^2 + 2 \langle I_N \rangle^2 + 4 I_i \langle I_N \rangle. \qquad (6.21)$$

Accordingly, from (6.18), (6.19) and (6.21) we have

$$N = [\langle I_N \rangle^2 + 2I_i \langle I_N \rangle]^{1/2},$$
$$= \langle I_N \rangle [1 + (2I_i/\langle I_N \rangle)]^{1/2}, \tag{6.22}$$

so that the signal-to-noise ratio is

$$I_i/N = I_i/\langle I_N \rangle [1 + (2I_i/\langle I_N \rangle)]^{1/2}. \tag{6.23}$$

In the limiting case when $I_i \gg \langle I_N \rangle$ (which is usually the situation for a hologram recording of good quality), the signal-to-noise ratio becomes

$$I_i/N = (I_i/2\langle I_N \rangle)^{1/2}. \tag{6.24}$$

With coherent illumination, the signal-to-noise ratio is proportional to the square root of the ratio of the intensities of the signal and the scattered background. Even a weak scattered background leads to relatively large fluctuations in intensity in the bright areas of the image.

7
Practical recording media

The ideal recording material for holography should have a spectral sensitivity well matched to available laser wavelengths, a linear transfer characteristic, high resolution and low noise. In addition, it should either be indefinitely recyclable or relatively inexpensive.

While several materials have been studied [Smith, 1977; Hariharan, 1980b], none has been found so far that meets all these requirements. However, a few have significant advantages for particular applications. This chapter reviews, in the light of the general considerations discussed in Chapter 6, the properties of some of these materials (see Table 7.1 for a summary of their principal characteristics).

7.1. Silver halide photographic emulsions

Silver halide photographic emulsions are still the most widely used recording material for holography, mainly because of their relatively high sensitivity and because they are commercially available. In addition, they can be dye sensitized so that their spectral sensitivity matches the most commonly used laser wavelengths.

An apparent drawback of photographic materials is that they need wet processing and drying; however, development is actually an amplification process, with a gain of the order of 10^6, which yields high sensitivity as well as a stable hologram. Another advantage of the formation of a latent image is that the optical properties of the recording medium do not change during the exposure, unlike materials in which the image is formed in real time. This makes it possible to record several holograms in the same photographic emulsion without any interaction between them.

Data on some of the silver halide photographic emulsions available for holography are summarized in Table 7.2, while typical $|t|$ *versus* E and spectral sensitivity curves are presented in figs. 7.1 and 7.2. Most of these emulsions are available coated on glass plates or film in a range of sizes and normally have an antihalation backing. Plates without any antihalation backing are available for making reflection holograms, though it is also possible to remove the antihalation backing with alcohol.

Table 7.1. *Recording materials for holography*

Material	Reusable	Processing	Type of hologram	Exposure required J/m^2	Spectral sensitivity nm	Resolution limit mm^{-1}	Max. diffraction efficiency (sine grating)
Photographic emulsions	No	Wet chemical	Amplitude (normal) Phase (bleached)	5×10^{-3}–5×10^{-1}	400–700	1000–10000	0.05 0.60
Dichromated gelatin	No	Wet chemical	Phase	10^2	350–580	>10000	0.90
Photoresists	No	Wet chemical	Phase	10^2	uv–500	3000	0.30
Photopolymers	No	Post exposure	Phase	10–10^4	uv–650	200–1500	0.90
Photochromics	Yes	None	Amplitude	10^2–10^3	300–700	>5000	0.02
Photothermoplastics	Yes	Charge and heat	Phase	10^{-1}	400–650	500–1200 (bandpass)	0.30
Photorefractive							
LiNbO$_3$	Yes	None	Phase	10^4	350–500	>1500	0.20
Bi$_{12}$SiO$_{20}$	Yes	None	Phase	10	350–550	>10000	0.25

Table 7.2. *Photographic materials for holography*

Plates	Type Film	Spectral sensitivity	Exposure (J/m^2) for $t=0.5$	Emulsion thickness μm Plate	Film	Grain size nm	Resolution limit mm^{-1}
Agfa-Gevaert (Holotest)							
10E56	10E56	Blue–green	10^{-2}	7	5	90	3000
10E75	10E75	Red	5×10^{-3}	7	5	90	3000
8E56HD	8E56HD	Blue–green	2.5×10^{-1}	7	5	35	> 3000
8E75HD	8E75HD	Red	10^{-1}	7	5	35	> 3000
Eastman Kodak							
649F	649F	Panchro	5×10^{-1}	15	6	60	> 3000
120-01/02	SO-173	Red	2×10^{-1}	6	6	50	> 3000
125-01/02	SO-424	Blue–green	5×10^{-2}	7	3	65	1250
131-01/02	SO-253	Red	3×10^{-3}	9	9	70	1250

Holograms recorded on these emulsions are processed using techniques similar to those used for normal photographic materials. Processing should be carried out immediately after exposure, since such fine-grain emulsions exhibit significant fading of the latent image.

Development is carried out in an active developer such as Kodak D-19 or Agfa 80 for about 5 min at 20°C, followed by a short rinse in water and fixing in either Kodak Rapid Fixer or Agfa-Gevaert G 334c for 3 min. Extending the fixing time serves no useful purpose and is liable to result in a lower optical density. The use of a stop bath is optional. To avoid reticulation, the temperatures of successive baths should not differ by more than $\pm 0.5°C$.

The processed plates are washed in running water for 10–15 min. A filter should be used on the water supply line to eliminate sediments, and a few

Fig. 7.1. Curves of amplitude transmittance $|t|$ against exposure (E) for typical photographic emulsions used for holography.

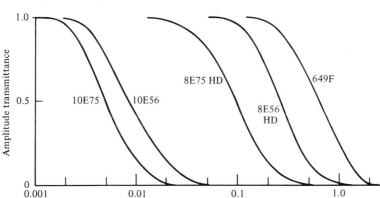

Fig. 7.2. Spectral sensitivity curves for typical photographic emulsions used for holography.

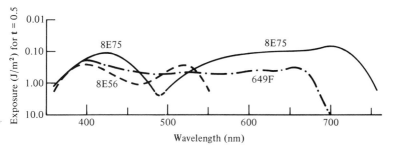

drops of a wetting agent should be added to the final rinse. The plates can then be stood on edge to dry at room temperature. Drying can be speeded up, if necessary, by soaking the plate in successive baths of denatured alcohol or isopropanol ($CH_3CHOHCH_3$), and blowing off the excess with a jet of dry air.

Maximum diffraction efficiency is usually obtained at an average amplitude transmittance of about 0.45 (corresponding to an optical density of 0.7, see Appendix A5) for transmission holograms, and the exposure should be adjusted accordingly. For reflection holograms, maximum diffraction efficiency is obtained at much higher optical densities, around 2.0.

7.1.1.   Emulsion shrinkage

The thickness of the processed photographic emulsion layer is usually about 15 per cent less than its original thickness, mainly because of the removal of the unexposed silver halide grains during fixing.

Because of this reduction in thickness, Λ' the fringe spacing in the hologram is less than Λ, the fringe spacing in the original interference pattern, except for the special case when the fringe planes are normal to the surface of the photographic emulsion (see section 4.11).

The effects of emulsion shrinkage are most noticeable in volume reflection holograms, since the fringe planes run almost parallel to the surface of the emulsion. When the hologram is illuminated with white light, it reconstructs an image at a wavelength λ' which is given by the relation

$$\lambda' = (\Lambda'/\Lambda)\lambda, \tag{7.1}$$

where λ is the wavelength used to record the hologram. Typically, a reflection hologram recorded with a He–Ne laser ($\lambda = 633$ nm) reconstructs at a wavelength $\lambda' \approx 530$ nm, giving a green image.

Because of emulsion shrinkage, amplitude holograms recorded in photographic emulsions also exhibit phase modulation which can modify the MTF. This phase modulation is mainly due to a surface relief structure arising from local tanning (hardening) of the gelatin by the oxidation products of the developer in the immediate vicinity of the reduced silver [Smith, 1968].

As shown in fig. 7.3, after processing, the entire emulsion layer is swollen to more than five times its normal thickness and is very soft. The tanned gelatin in the vicinity of a clump of developed silver grains absorbs less water than the untanned areas and therefore dries more quickly. Shrinkage during drying pulls some of the gelatin from adjacent untanned areas into the image areas, and, when the emulsion has dried, the higher density areas stand out in relief because both the image silver and the extra gelatin contribute to their bulk.

Since the relief image formed by local tanning is usually confined to low spatial frequencies (< 200 mm^{-1}), it normally contributes very little to the reconstructed image. However, it can give rise to intermodulation noise. The formation of such a relief image can be avoided by the use of a developer with a high sulphite content. Alternatively, the emulsion can be treated in a prehardening bath before processing.

7.1.2.   Modulation transfer function

The MTF curves for the fine-grain photographic materials used for holography usually extend to quite high spatial frequencies. While the MTF can be measured with specially designed instruments [Biedermann & Johansson, 1972; 1975] it is derived most commonly in an indirect fashion from measurements of the diffraction efficiency of sinusoidal gratings with differential spatial frequencies recorded on these materials [Buschmann, 1972]. In the latter case, these values must be corrected for losses due to Fresnel reflection at the surfaces of the emulsion layer.

Typical curves of diffraction efficiency as a function of spatial frequency obtained for 8E75, 8E56, 10E75 and 10E56 emulsions for different wavelengths are presented in fig. 7.4. As can be seen, after an initial rapid drop, all of them exhibit a gently sloping region covering the range of spatial frequencies used in holography. However, beyond a certain limit, the diffraction efficiency again decreases rapidly with increasing spatial frequency. These effects are more pronounced for the more sensitive emulsions (10E75 and 10E56) which have larger grains, and can be

Fig. 7.3. Formation of a relief image due to local tanning of the gelatin [Smith, 1968].

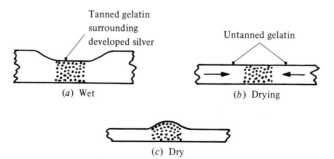

Tanned gelatin
surrounding
developed silver

Untanned gelatin

(*a*) Wet

(*b*) Drying

(*c*) Dry

attributed to scattering in the virgin emulsion during the exposure [Buschmann, 1972; Joly & Vanhorebeek, 1980].

7.1.3. Scattered flux spectra

Figure 7.5 shows normalized scattered flux spectra for three typical photographic emulsions used for holography [Biedermann, 1970] exposed uniformly and processed to an amplitude transmittance $t \approx 0.45$ (optical density, $D \approx 0.7$).

These curves show that the scattered flux is higher for the most sensitive emulsion (10E70, an earlier version of 10E75) which has larger grains; in addition, exposure to laser light results in higher scattering than exposure to an incoherent source of the same wavelength. This is due to the speckle pattern produced in the emulsion layer during the exposure by coherent light scattered from individual grains. As expected, the section of the scattered flux spectrum extending to higher spatial frequencies, which is due to the grains, is very nearly a straight line; however, there is a steep rise at lower spatial frequencies due to the relief image formed in the gelatin.

Fig. 7.4. MTF curves for typical photographic emulsions used for holography [Buschmann & Metz, 1971].

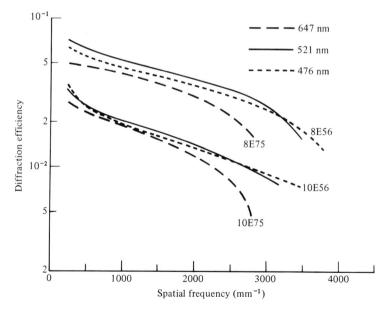

7.1.4. Monobath processing

Processing of holograms recorded on photographic emulsions for some purposes such as real-time hologram interferometry (see section 14.1) can be simplified and speeded up by the use of a monobath in which development and fixing take place simultaneously [Ragnarsson, 1970; Hariharan, Ramanathan & Kaushik, 1973].

The formula for a monobath which can be used with 10E75 plates is shown in Table 7.3. It is basically an active developer formula to which a silver halide complexing agent has been added. The image is formed initially by chemical development of the exposed silver halide grains. At the same time, some of the unexposed silver halide dissolved by the complexing agent migrates towards adjacent developing grains and is reduced there to metallic silver (physical development). As a result, there is an increase in speed of about 50 per cent with little change in the brightness of the reconstructed image.

A problem with monobaths is their tendency to form a sludge of metallic silver after use; this can be avoided either by discarding the monobath immediately after use or by using a sequestering agent [Dietrich, Raine & O'Brien, 1976].

Fig. 7.5. Normalized scattered flux spectra for typical photographic emulsions used in holography [Biedermann, 1970].

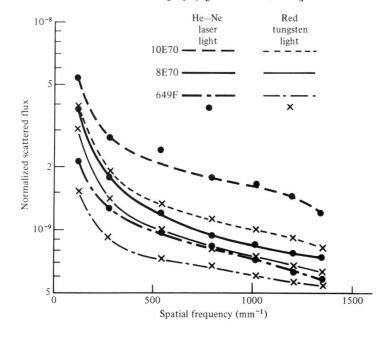

7.1.5. Bleach techniques

Because of the relatively low diffraction efficiency of amplitude holograms, holograms recorded in photographic emulsions are often processed to yield phase holograms, which have much higher diffraction efficiencies.

Two basic processes can be envisaged. In one, the variations in the optical density of the hologram are converted into thickness variations of the emulsion layer [Cathey, 1965; Altman, 1966]. In the other, the variations in the optical density are converted into a local modulation of the refractive index [Burckhardt, 1967]. Because the first method is effective only at relatively low spatial frequencies in the case of photographic materials and also has a lower maximum diffraction efficiency (0.339 for a surface relief grating against 1.00 for a volume phase grating), photographic phase holograms are almost exclusively of the latter type.

Early work on photographic volume phase holograms was carried out with bleaches which converted the image silver in the developed and fixed photographic emulsion into a transparent silver salt with a higher refractive index than gelatin, as shown in fig. 7.6(*a*). Bleach baths used for this purpose contain a strong oxidizing agent such as mercuric chloride, cupric bromide, ferric bromide, potassium dichromate or potassium ferricyanide [Upatnieks & Leonard, 1969]. Alternatively, it is possible to bleach the hologram after drying, using bromine vapour [Graube, 1974].

While volume phase holograms made with such bleaches have good diffraction efficiencies, they exhibit quite high levels of scattered light. This is partly due to nonlinear effects arising from the formation of a relief image

Table 7.3. *Monobath formula for 10E75 plates* [*Hariharan, Ramanathan &* *Kaushik, 1973*]

Stock solution A		
Metol	3.5	g
Sodium sulphite (anhydrous)	50	g
Hydroquinone	15	g
Potassium alum	5	g
Sodium thiosulphate	50	g
Distilled water	800	ml
Stock solution B		
Sodium hydroxide	10	g
Distilled water	200	ml

Mix 4 parts of A with 1 part of B before use.
Process plates for 2 min at $20 \pm 0.5\,°C$.

at low spatial frequencies [Upatnieks & Leonard, 1970] and partly due to scattering from the individual grains making up the recording. Since the scattering is proportional to the sixth power of the effective radius of the grains [Benton, 1971; Hariharan, Kaushik & Ramanathan, 1972] the size of the grains produced must be kept to a minimum. However, in this case, there is always considerable grain growth due to solution physical development; in addition, the tanning effect of the developer can also create a shell of hardened gelatin around each grain so that its effective size is increased [van Renesse, 1980].

For these reasons, recent workers have favoured reversal bleach processes in which, as shown in fig. 7.6(*b*), the fixing bath is omitted and the developed silver is dissolved away, leaving a phase hologram made up of the undeveloped silver halide crystals. These, being smaller, give rise to much less scattering [Chang & George, 1970; Hariharan, 1971; Buschmann, 1971]. This technique also has the advantage that the relief image arising from local tanning by the developer depresses the MTF at low spatial frequencies resulting in a reduction in intermodulation noise [Lamberts & Kurtz, 1971]. Figure 7.7 shows the ratio of the signal and noise powers for holograms of a diffusing object processed in conventional and reversal bleaches [Buschmann, 1971]. The use of a tanning developer as well as a tanning bleach also helps to minimize emulsion shrinkage [Phillips, Ward, Cullen & Porter, 1980]. Formulas for such a developer and such a bleach are given in Table 7.4.

A major problem with bleached holograms is their tendency to darken

Fig. 7.6. Production of photographic volume phase holograms by (*a*) conventional and (*b*) reversal bleach techniques.

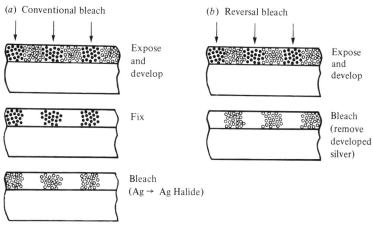

98 Practical recording media

Fig. 7.7. Curves showing the ratio of the signal and noise powers for conventional and reversal bleaches [Buschmann, 1971].

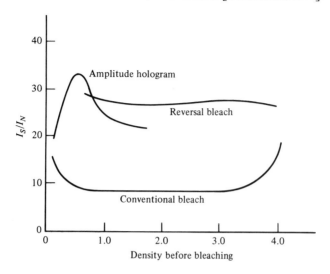

Table 7.4

Tanning developer GP 62		
Stock solution A		
Metol	15	g
Pyrogallol	7	g
Sodium sulphite	20	g
Potassium bromide	4	g
EDTA (sodium salt)	2	g
Distilled water	1000	ml
Stock solution B		
Sodium carbonate	60	g
Distilled water	1000	ml

Mix 1 part A, 1 part B and 2 parts water. Develop for 2 min at $20\pm0.5°$C. Rinse for 2 min, then bleach till clear.

Tanning bleach GP 432		
Potassium bromide	50	g
Boric acid	1.5	g
Distilled water	1000	ml

Add p-benzoquinone 2 g/l just before use.

when exposed to ambient light because of the formation of printout silver. This can be minimized by treatment with a desensitizing dye [Buschmann, 1971], but this has the drawback that it stains the gelatin. However, the resistance of the hologram to printout also depends on the silver salt of which it is formed, AgCl being the poorest in this respect, while AgBr is considerably better and AgI is extremely stable [McMahon & Maloney, 1970]. It is possible to convert a hologram recording consisting of AgCl or AgBr to AgI by soaking it in a dilute solution of KI [Hariharan & Ramanathan, 1971]. An even simpler method is to use a reversal bleach containing a small amount of KI (see Table 7.5) [Hariharan, Ramanathan & Kaushik, 1971]. Comparative measurements of stability against exposure to light of untreated and treated holograms are presented in fig. 7.8.

7.2.   Dichromated gelatin

Dichromated gelatin is, in some respects, an almost ideal recording material for volume phase holograms; it has large refractive-index-modulation capability, high resolution and low absorption and scattering.

Hologram recording in dichromated gelatin makes use of the fact that a gelatin layer containing a small amount of a dichromate such as $(NH_4)_2Cr_2O_7$, becomes progressively harder on exposure to light. This hardening is due to the photochemically produced Cr^{3+} ion forming localized cross-links between the carboxylate groups of neighbouring gelatin chains.

This effect has been used in the past to produce a relief image by washing off the unhardened gelatin with warm water. However, much better phase

---

Table 7.5. *Reversal bleach for maximum stability* [*Hariharan, Ramanathan &
Kaushik, 1971*]

===

Stock solution A
  Potassium dichromate   8 g
  Conc. sulphuric acid   10 ml
  Distilled water   1000 ml

Stock solution B
  Potassium iodide   2 g
  Distilled water   1000 ml

Mix 1 part A, 1 part B and 8 parts water.
Bleach for 5 min at 20°C. Use only once.

===

holograms can be obtained if the gelatin film is processed to obtain a modulation of the refractive index [Shankoff, 1968; Lin, 1969].

7.2.1.   Film preparation and processing

Since plates and films coated with dichromated gelatin are not commercially available, it is necessary to prepare them in the laboratory. It is possible to coat glass plates with a uniform gelatin film [Brandes, Francois & Shankoff, 1969; McCauley, Simpson & Murbach, 1973]. In this case it is necessary to harden the gelatin film initially to a degree where it will not dissolve in water during subsequent processing, but still remains soft enough for the photochemical reaction to produce a significant difference in the local hardness. Alternatively, a method which has been used widely is to dissolve out the silver halide in the emulsion layer in a photographic plate. Kodak 649F plates have been used commonly because they have a fairly thick (≈ 15 μm) coating of gelatin, which is not excessively hardened. A suitable processing schedule is given in Table 7.6 [Chang & Leonard, 1979].

To sensitize the plates, they are soaked for about 5 min at 20°C in an aqueous solution of $(NH_4)_2Cr_2O_7$ (≈ 5 per cent) to which a small amount of a wetting agent has been added. They are then allowed to drain on edge and dried at 25–30°C in darkness or under a red safelight. Since the Cr^{6+} ion is reduced slowly to Cr^{3+} even in the dark, the plates should be used within a well defined period after they are sensitized (usually 8–24 hours) to ensure reproducible results.

Fig. 7.8. Increase in optical density of bleached holograms when exposed to noon sunlight: (a) bleach containing KBr, (b) bleach followed by treatment in 0.2 per cent KI, (c) bleach containing 0.2 per cent KI [Hariharan & Ramanathan, 1971].

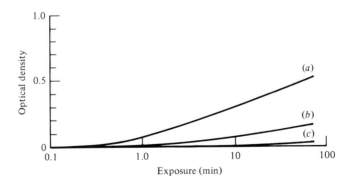

After exposure in the holographic setup using blue light from an Ar^+ laser ($\lambda = 488$ nm, $E \approx 10^2$ J/m^2), the gelatin film is soaked in a 0.5 per cent solution of ammonium dichromate as well as in an additional hardening bath to control the overall hardness of the film. The film is then washed in water at 20–30°C for 10 min to remove the dichromate. Since the melting

Table 7.6. *Processing schedule for dichromated gelatin holograms recorded on Kodak 649 F plates* [*Chang & Leonard, 1979*]. (*All solutions at 20°C unless indicated otherwise.*)

(a) Preparation of plates

		Lighting	Time
1.	Soak in Kodak fixer (without hardener)	Normal	10 min
2.	Wash in running water (raise temperature slowly from 20 to 33°C, then lower slowly to 20°C)		5 min
3.	Soak in Kodak fixer with 3.25 per cent hardener		10 min
4.	Wash in running water		10 min
5.	Rinse in distilled water		5 min
6.	Rinse in Photo-Flo solution (1 drop/500 ml)		30 s
7.	Drain, wipe off excess and dry at room temperature		
8.	Soak in 5 per cent ammonium dichromate solution with 0.5 per cent Photo-Flo	Red	5 min
9.	Drain, wipe glass side and dry at room temperature		

(b) Development of plates

		Lighting	Time
1.	Soak in 0.5 per cent ammonium dichromate solution	Red	5 min
2.	Soak in Kodak fixer with 3.25 per cent hardener		5 min
3.	Wash in running water	Normal	10 min
4.	Rinse in Photo-Flo solution (1 drop/500 ml) and wipe		30 s
5.	Soak in distilled water		2 min
6.	Dehydrate in 50/50 solution of distilled water and isopropanol		3 min
7.	Dehydrate in 100 per cent isopropanol		3 min
8.	Dry (room temperature, low humidity)		1 h
9.	Bake or evaporate residual water under vacuum		2 h

point of the gelatin film is higher than the temperature of the water, none of the gelatin goes into solution. However, the film absorbs a large amount of water and swells. Since the rigidity of the gel is higher in the hardened areas, these swell to a lesser extent than the unhardened areas.

The crucial stage in processing is the next step. In this, the swollen film is immersed for 3 min in two successive baths of isopropanol at 20–30°C and then dried thoroughly. To avoid condensation of moisture, which can result in reduced diffraction efficiency, some authors, such as Meyerhofer [1972], recommend the use of a blast of warm air as the film is pulled out of the second bath. However, too rapid drying can result in a milky-white film which scatters strongly.

Two successive baths of isopropanol are necessary to ensure that the final bath is free of water, and dehydration is complete. Raising the temperature of the water and isopropanol baths increases the sensitivity of the film, but can also result in an increase in scattering. Any residual water is eliminated by drying the plates in a vacuum for 1–2 hours. Since moisture, or high ambient humidity, rapidly degrades holograms recorded on dichromated gelatin, they should be protected by a cover glass cemented on with an epoxy cement.

If the sensitizing and processing sequence is carried out properly, it should result in a hologram with high diffraction efficiency and low scattering. However, if the results are not satisfactory, it is possible, within limits, to reprocess the hologram and obtain a better result [Chang, 1976]. With reflection holograms, the wavelength at which the hologram reconstructs an image can be controlled by regulating the degree of swelling of the gelatin [Coleman & Magariños, 1981].

A critical factor in obtaining good results with dichromated gelatin is the degree of preliminary hardening of the gelatin layer. If it is too low, the resulting holograms are noisy, while if it is too high, the available refractive index modulation and the sensitivity decrease. The drop in sensitivity when the gelatin layer is hardened, prior to exposure, with a chromium salt is because many of the carboxyl groups have already reacted and are no longer available. This can be avoided if a noble metal salt ($H_2PtCl_6 \cdot 6H_2O$) or formaldehyde, both of which form cross-links with the amino groups and not with the carboxyl groups, is used to preharden the gelatin [Mazakova, Pancheva, Kandilarov & Sharlandjiev, 1982b]. An increase in sensitivity by a factor of 10 can be obtained by this technique.

7.2.2.   Resolution and spectral sensitivity

The characteristics of dichromated gelatin as a holographic

recording medium have been discussed in some detail by Chang [1979]. Dichromated gelatin has excellent resolution and an MTF which is almost flat out to a spatial frequency of 5000 mm^{-1}. It is therefore very suitable for making reflection holograms as well as transmission holograms.

Very large refractive-index modulation can be obtained with dichromated gelatin layers; values as large as 0.08, with very low scattering, are possible with careful processing [Chang & Leonard, 1979; Salminen & Keinonen, 1982]. Typical curves showing the diffraction efficiencies of transmission and reflection gratings recorded in dichromated gelatin are presented in fig. 7.9. These show that diffraction efficiencies close to the theoretical maximum for volume holograms can be easily obtained.

The spectral sensitivity of dichromated gelatin at 514 nm is normally only about a fifth of that at 488 nm and it drops to zero at around 580 nm. The possibility of extending it to longer wavelengths for holographic recording was first explored by Graube [1973], who used dichromated gelatin sensitized with triphenylmethane dyes to record holograms with a He–Ne laser. Good diffraction efficiency can be obtained with exposures of 500–1000 J/m^2 at 633 nm if methylene blue in an aqueous dichromate solution is used as a sensitizer [Kubota & Ose, 1979a].

7.2.3. Mechanisms of image formation

The chemical reactions and physical mechanisms involved in

Fig. 7.9. Diffraction efficiency of gratings recorded in dichromated gelatin as a function of exposure [Chang, 1979].

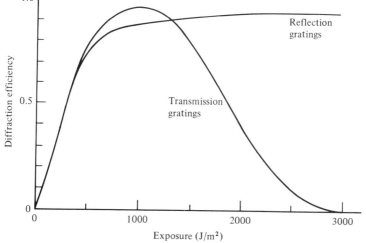

recording a volume hologram in a film of dichromated gelatin are still not fully understood.

The earliest explanation of the large local changes in refractive index observed was put forward by Curran & Shankoff [1970], who suggested that cracks were formed at the fringe boundaries and the dimensions of these cracks increased with increasing exposure. While the presence of voids within the gelatin was confirmed by the drop in diffraction efficiency when the hologram was immersed in an index-matching fluid, this explanation was not entirely satisfactory since such cracks would scatter light quite strongly. An alternative theory proposed by Meyerhofer [1972], in which the index modulation was caused by the bonding of isopropanol molecules to the chromium atoms at the sites of the cross-links, is contradicted by the fact that the hologram image is destroyed by very small amounts of water.

Perhaps the best explanation of the excellent optical properties of volume holograms recorded in dichromated gelatin is due to Case & Alferness [1976] who suggested that instead of cracks, a very large number of small vacuoles, with dimensions much smaller than the wavelength of light, were formed in the unhardened areas. This would allow a smooth variation in the refractive index across a fringe and account for the low scattering. Calculations of the modulation of the refractive index and the diffraction efficiency based on their model are in good agreement with experimental observations. In addition, an overall decrease in the refractive index, predicted by their model, has been found. It appears [Chang & Leonard, 1979; Graube, personal communication] that hologram formation in dichromated gelatin takes place in two stages, involving rearrangement of the gelatin chains as well as the formation of voids. However, work by Sjolinder [1981] suggests that the observed refractive index modulation is at least partly due to the formation in the hardened areas, of complexes with a high refractive index involving a Cr^{3+} compound, gelatin and isopropanol.

7.2.4.   Silver-halide sensitized gelatin

An alternative technique, which combines the relatively high sensitivity of photographic materials with the low scattering and high light-stability of dichromated gelatin, involves exposing a silver halide photographic emulsion and then processing it so as to obtain a volume phase hologram consisting solely of hardened gelatin.

This technique which was originally developed by Pennington, Harper & Laming [1971] has been studied in more detail by Graver, Gladden & Estes

[1980] and by Chang & Winick [1980]. As shown in the processing schedule in Table 7.7, the exposed photographic emulsion is developed in a non-tanning developer and then bleached in a bath containing ammonium dichromate. During the bleaching process, the developed silver is oxidized to Ag^+, while the Cr^{6+} ion in the bleach is reduced to the Cr^{3+} ion. These Cr^{3+} ions form cross-links between the gelatin chains in the vicinity of the oxidized silver grains causing local hardening. The emulsion is then fixed to remove the unexposed silver halide, washed and dehydrated with isopropanol and dried exactly as for a dichromated gelatin hologram. Transmission gratings with a diffraction efficiency ≈ 70 per cent have been produced with Kodak 649F plates.

Table 7.7. *Processing schedule for silver-halide sensitized gelatin holograms* (*Kodak 649F plates*) [*Chang & Winick, 1980*] (*All solutions at 22°C*)

Step		Lighting
1.	Develop in Kodak D-19 developer for 5 min	None
2.	Soak in stop bath for 30 s	
3.	Rinse for 15 s	
4.	Bleach for 150 s (for formula see below)	Red
5.	Rinse for 15 s	
6.	Soak in 0.5 per cent $(NH_4)_2Cr_2O_7$ for 5 min	
7.	Soak in hardening fixer for 5 min	
8.	Wash in running water for 10 min	Normal
9.	Wash in distilled water for 3 min	
10.	Dehydrate in 50 per cent isopropanol for 3 min	
11.	Dehydrate in 100 per cent isopropanol for 3 min	
12.	Dry in air and then in a vacuum chamber	

Bleach formula
Solution A

Distilled water	500 ml
Ammonium dichromate	20 g
Concentrated sulphuric acid	14 ml
Distilled water to make	1000 ml

Solution B

Potassium bromide	92 g
Distilled water to make	1000 ml

Just before use, mix one part A with ten parts distilled water, then add one part B.

The resolution possible with this technique is limited by the grain size of the photographic emulsion. Higher resolution can be attained with specially prepared gelatin layers [Mazakova, Pancheva, Kandilarov & Sharlandjiev, 1982a]. For this, glass plates coated with a 20 μm thick layer of gelatin containing 1 per cent of $(CH_3COO)_3Cr$ as a hardener are soaked in a solution of $AgNO_3$ and bathed in an alkali halide solution so that very fine grains of silver halide are formed in the gelatin layer. The excess alkali halide is then washed away and the layer is sensitized by treating it with a solution of dichromate and an optical sensitizer.

When exposed to light from an Ar^+ laser ($\lambda = 488$ nm) and processed in the same manner as dichromated gelatin, reflection holograms with a diffraction efficiency of 50 per cent and very low scattering are produced. The sensitivity is about 10 times higher than that of dichromated gelatin.

7.3.   Photoresists

Photoresists are light-sensitive organic films which yield a relief image after exposure and development. Several photoresists have been used to record holograms [Bartolini, 1977a], but they are all relatively slow, typically requiring an exposure of 10^2 J/m^2 to blue light ($\lambda = 442$ μm), and since a thin phase hologram is formed, nonlinear effects are noticeable at diffraction efficiencies greater than about 0.05. However, they have the advantage that replication using thermoplastics is easy.

Two types of photoresists are available. In negative photoresists, the exposed areas become insoluble and the unexposed areas are dissolved away during development. Relatively long exposures are necessary with such photoresists, usually through the back of the plate, to ensure that the exposed photoresist adheres to the substrate during development. Because of this, positive photoresists in which the exposed areas become soluble and are washed away during development are preferable.

The most widely studied of these is Shipley AZ-1350. The sensitivity of this photoresist is a maximum in the ultraviolet and drops rapidly for longer wavelengths towards the blue. Hence, holograms are best recorded either with an Ar^+ laser at 458 nm, or with a He–Cd laser at 442 nm. Even though the power available at these wavelengths is much less than that available from the Ar^+ laser at 488 nm, this is more than compensated for by the increase in sensitivity [Beesley, Castledine & Cooper, 1969].

The photoresist is usually coated on a glass substrate by spinning to give a layer from 1 to 2 μm thick. This is then baked at 75°C for 15 min to ensure complete removal of the solvent. As shown in fig. 7.10, an approximately linear depth *versus* exposure characteristic is obtained if AZ-303 A developer, an alkaline solution, is used instead of the normal developer for this material. In addition, AZ-303 A developer gives an increase in sensitivity by a factor of 2 to 3 over normal processing. To control the etch rate, the AZ-303 A developer is normally diluted with four parts of distilled water. The exposure can also be reduced significantly, and improved linearity obtained, by uniformly pre- or post-exposing the resist to an incoherent source such as a fluorescent lamp [Bartolini, 1972, 1974; Norman & Singh, 1975; Livanos, Katzir, Shellan & Yariv, 1977].

The MTF curve for AZ-1350 photoresist is almost flat up to spatial frequencies of 1500 mm^{-1}, and does not appear to be affected significantly by small changes in the coating thickness. In addition, since the material has no grain structure, scattered light is negligible. These characteristics, as well as the convenience of replication have led to the widespread use of AZ-1350 photoresist in the production of blazed holographic diffraction gratings (see section 12.9).

7.4.   Photopolymers

A number of organic materials are known which can be activated through a photosensitizer to exhibit thickness and refractive index changes due to photopolymerization or cross-linking. Thick layers can be made to yield volume phase holograms with high diffraction efficiency and high angular selectivity, which can be viewed immediately after exposure. After

Fig. 7.10. Depth *versus* exposure characteristics obtained with AZ-1350 photoresist [Bartolini, 1974].

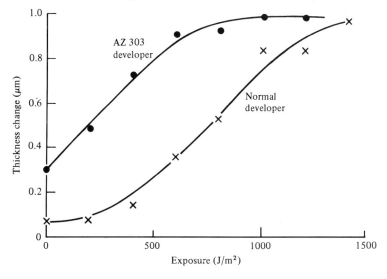

the exposure a continuing dark reaction due to diffusion of the monomer into the zones of polymerization increases the refractive index modulation. The hologram is then stabilized by a uniform post-exposure to complete the reaction. The characteristics of most of these materials have been reviewed by Bartolini [1977*b*]. In addition, Booth [1975, 1977] has discussed the characteristics and possible applications of a photopolymer which is commercially available.

Photopolymers have still not found wide use in holography; this is mainly because of their low sensitivity and relatively short shelf life.

7.5.   Photochromics

Photochromic materials undergo reversible changes in colour when exposed to light. Several organic photochromics have been studied [Bartolini, 1977*b*], but they are prone to fatigue and have a limited life. Inorganic photochromics are usually crystals doped with selected impurities: photochromism is due to reversible charge transfer between two species of electron traps. A detailed survey of such materials has been made by Duncan & Staebler [1977].

Photodichroic crystals have recently received more attention. In these, the absorption centres are anisotropic and selective alignment of the anisotropic absorption centres is induced by exposure to linearly polarized light. The result is a preferred orientation for absorption, which can be switched between two orthogonal directions. These crystals have a higher sensitivity (typical exposure 10^2 J/m^2) than photochromic crystals (typical exposure 10^3 J/m^2). In addition, a single laser can be used for storage, readout and erasure; only the direction of linear polarization is changed. The most commonly used photodichroic crystals are alkali halides; their characteristics have been reviewed by Casasent & Caimi [1977].

Inorganic photochromics are grain free and have high resolution. Because of their relatively large thickness, a number of holograms can be stored in them. In addition, they require no processing and can be erased and re-used almost indefinitely. However, despite these advantages, their use has been limited so far by their low diffraction efficiency (<0.02) and their low sensitivity.

7.6.   Photothermoplastics

A surface relief hologram can also be recorded in a thin layer of a thermoplastic. For this, it is combined with a photoconductor and charged to a high voltage. On exposure, a spatially varying electrostatic field is created. The thermoplastic is then heated so that it becomes soft enough to be deformed by the field and finally cooled to fix the pattern of deformations [Urbach, 1977].

Such materials have a reasonably high sensitivity over the whole visible spectrum and yield a thin phase hologram with fairly high diffraction efficiency. In addition, they have the advantage that they do not require wet processing. If a glass substrate is used, the hologram can be erased and the material re-used a number of times. Hologram recording cameras using photothermoplastics are extremely useful for applications such as hologram interferometry (see Chapters 14 and 15) and optical information processing (see Chapter 13) [Ineichen, Liegeois & Meyrueis, 1982].

The most widely used type of photothermoplastic is a multilayer structure consisting of a substrate (glass or Mylar) coated with a thin, transparent, conducting layer (usually indium oxide), a photoconductor and a thermoplastic [Urbach & Meier, 1966; Lin & Beauchamp, 1970*b*].

As shown in fig. 7.11, the film is initially sensitized in darkness by applying a uniform electric charge to the top surface. This is done by a corona device which moves over the surface at a constant distance from it and sprays positive ions on to it. As a result, a uniform negative charge is induced on the conductive coating on the substrate.

Fig. 7.11. Record-erase cycle for a photothermoplastic recording material [Lin & Beauchamp, 1970*b*].

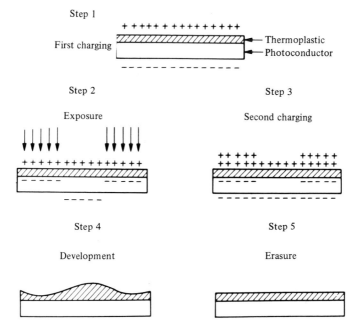

In the second step, when the film is exposed, charge carriers are produced in the photoconductor wherever light is incident on it. These migrate to the two oppositely charged surfaces and neutralize part of the charge deposited there during the sensitizing step. This reduces the surface potential but does not change the surface charge density and the electric field, so that the image is not developable at this stage.

Accordingly, in the next step, the surface is recharged once again to a constant potential using the same procedure as in the first step. As a result, additional charges are deposited wherever the exposure had resulted in a migration of charge. The electric field now increases in these regions, producing a spatially varying field pattern and hence a developable latent image.

In the fourth step, this latent image is developed by heating the thermoplastic uniformly to a temperature near its softening point. This is done most conveniently by passing a current briefly through the conductive coating on the substrate. The thermoplastic layer then undergoes local deformation as a result of the varying electric field across it, becoming thinner wherever the field is higher (the illuminated areas) and thicker in the unexposed areas. Once the thermoplastic layer has cooled to room temperature, this thickness variation is frozen in, so that the hologram is quite stable.

Since, after the second charging, the latent image is relatively insensitive to exposure to light, it is possible to monitor the diffraction efficiency of the hologram during development and terminate the application of heat at the proper time.

Finally, when the plate is to be re-used, it is flooded with light and the thermoplastic layer is heated to a temperature somewhat higher than that used for development. At this temperature the thermoplastic is soft enough for surface tension to smooth out the thickness variations and erase the previously recorded hologram. A blast of compressed air or dry nitrogen is then used to cool the photothermoplastic rapidly to room temperature, in preparation for the next exposure.

An alternative method of development which has been studied by Saito, Imamura, Honda & Tsujiuchi [1980] and which shows promise is the use of solvent vapours to soften the thermoplastic. This has the advantage that it eliminates the need to heat the substrate. In addition it gives higher sensitivity and lower noise. Enhanced sensitivity can also be obtained by the use of double-layer and triple-layer photoconductor systems [Saito, Imamura, Honda & Tsujiuchi, 1981].

7.6.1. Materials

The most commonly used photoconductor is poly-*n*-vinyl carbazole (PVK) sensitized with 2,4,7 trinitro-9-fluorenone (TNF). This material has good sensitivity but a problem with it is ghost images – the reappearance of a previously erased image in a subsequent processing cycle. This has been shown to be due mainly to low electron mobility [Colburn & Dubow, 1973] and can be minimized by increasing the TNF/PVK ratio and flooding the material with light prior to or during erasure.

The life of these materials is limited mainly by the thermoplastic layer. The material used by most early experimenters was Staybelite Ester 10 (Hercules Inc., Wilmington, Delaware, U.S.A.). However, while Staybelite is otherwise satisfactory, it cannot be recycled more than 10–100 times before its properties are degraded significantly. The main cause of this degradation is ozone formed during the charging process. When protected from degradation by ozone, other materials such as a styrene-methacrylate copolymer appear capable of extended life with good spatial-frequency response and diffraction efficiency. Currently, commercially available photothermoplastics are reported as having a life of more than 300 cycles.

7.6.2. Response

The MTF of all photothermoplastics is limited to a band of spatial frequencies. The spatial frequency corresponding to the peak response depends mainly on the thickness of the thermoplastic layer, but the usable bandwidth is determined by a number of parameters, including the properties of the material used, the field strength and the manner of development. Commercial photothermoplastics have a peak response at spatial frequencies around 800 mm^{-1}.

A typical curve showing the diffraction efficiency obtained, as a function of spatial frequency, for gratings recorded with two plane wavefronts of equal amplitude is shown in fig. 7.12, while fig. 7.13 shows the variation of the diffraction efficiency with exposure for a grating with a spatial frequency of 800 mm^{-1}. A diffraction efficiency close to the theoretical maximum of 0.339 for a thin phase hologram can be obtained with an exposure of less than 0.1 J/m^2.

7.6.3. Noise

A problem encountered with most photothermoplastics is 'frost' – a random surface modulation whose power spectrum is very similar to the MTF of the material. The exact causes of frost are still not known, but it can be minimized by using thinner thermoplastic films and lower charging voltages. It has also been shown [Reich, Rav-Noy & Friesem, 1977] that frost is reduced with a thermoplastic having a narrow distribution of molecular weights. In addition, the use of a thin, electrically insulating layer between the photoconductor and thermoplastic layers results in a substantial reduction in frost [Lee, Lin & Tufte, 1977].

Intermodulation noise would normally be expected to be a problem in any thin phase hologram. A rather surprising feature of the image reconstructed by a hologram recorded on a photothermoplastic is, therefore, the relative lack of intermodulation noise even at quite high diffraction efficiencies [Credelle & Spong, 1972]. This is because intermodulation noise arises from terms in (6.14) whose spectra involve

Fig. 7.12. Diffraction efficiency as a function of spatial frequency for gratings recorded in a typical photothermoplastic.

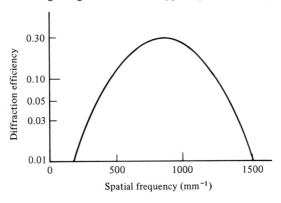

Fig. 7.13. Diffraction efficiency as a function of exposure for gratings recorded in a typical photothermoplastic.

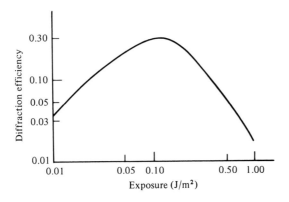

multiple correlations and convolutions of the object spectrum and are therefore mainly confined to low spatial frequencies. Since the response of the photothermoplastic at such low spatial frequencies is virtually nil, these terms are severely attenuated.

7.7. Photorefractive crystals

In some electro-optic crystals, exposure to light frees trapped electrons, which then migrate through the crystal lattice and are again trapped in adjacent unexposed regions. This migration usually occurs through diffusion or an internal photovoltaic effect. The spatially varying electric field produced by the resulting space-charge pattern modulates the refractive index through the electro-optic effect, resulting in the formation of a phase hologram. When desired, this hologram can be erased by uniformly illuminating the crystal, which can thus be recycled almost indefinitely.

7.7.1. LiNbO$_3$

A number of photorefractive crystals have been studied for hologram recording [Staebler, 1977]. The most widely used of these is Fe-doped LiNbO$_3$. This gives quite high diffraction efficiency and, because of the high angular sensitivity of such a thick recording medium, it is possible to store a large number of holograms in a single crystal [Burke, Staebler, Phillips & Alphonse, 1978; Glass, 1978]. However, the high angular sensitivity of the hologram also creates problems. To satisfy the Bragg condition, it is normally necessary to use the same wavelength for readout as for recording, and this results in degradation of the stored image. This can be overcome by using light of a longer wavelength, to which the crystal is insensitive, for readout and satisfying the Bragg condition by making use of the crystal birefringence [Petrov, Stepanov & Kamshilin, 1979a, b]. Another major drawback which still remains, though, is the extremely low sensitivity of LiNbO$_3$ (typical exposure 10^4 J/m^2). Higher sensitivity can be obtained with Sr$_{0.75}$Ba$_{0.25}$Nb$_2$O$_6$ [Thaxter & Kestigian, 1974].

7.7.2. Bi$_{12}$SiO$_{20}$ and Bi$_{12}$GeO$_{20}$

Much higher sensitivity has been obtained with photoconductive electro-optic crystals such as Bi$_{12}$SiO$_{20}$ (BSO) and Bi$_{12}$GeO$_{20}$ (BGO) by the application of an external electric field. A device known as the Pockels Readout Optical Modulator (PROM) [Hou and Oliver, 1971] uses a thin crystal slice oriented normal to the 100 axis with transparent electrodes on its two faces. An electric field parallel to the 100 axis is applied to the crystal by means of these electrodes. When the crystal is exposed to light, photoconductivity results in decay of the stored field in the illuminated areas. Due to the Pockels effect, the crystal then develops fast and slow propagation modes polarized along the two axes lying in the plane of the slice, the differences in the refractive indices being proportional to the electric field.

While the sensitivity in this configuration is high (typically 0.5 J/m^2 at $\lambda = 440$ nm), the resolution is limited to about 500 mm^{-1} by photocarrier drift through the thickness of the crystal. This limitation has been eliminated by using a transverse field, the holographic fringes being perpendicular to the direction of the applied field as shown in fig. 7.14 [Huignard & Micheron, 1976]. The photosensitivity with this configuration is lower (3 J/m^2 for BSO, which is the better of the two, for a diffraction efficiency of 0.01), but still comparable to that for slow photographic materials. A maximum diffraction efficiency of 0.25 is possible with a field of 900 V/mm, and the storage time constant in darkness is about 30 h. Erasure is achieved by uniform illumination of the crystal, which can be recycled indefinitely without fatigue effects. Typical recording-erasure curves for BSO at different applied fields at an incident power density of 2.45 W/m^2 ($\lambda = 514$ nm) are presented in fig. 7.15. Experiments with a frequency-doubled Nd:YAG laser show that the response time depends on the incident power density and can be reduced to < 5 ns, provided the incident power density is high enough ($> 10^7$ W/m^2) [Hermann, Herriau & Huignard, 1981]. Higher diffraction efficiency and an increase in the holographic sensitivity can be obtained by preliminary infrared irradiation ($\lambda = 650$–800 nm) [Kamshilin & Miteva, 1981].

A feature of BSO, when used in this configuration, is that the form of the MTF depends on the applied field [Huignard, Herriau, Rivet & Günter, 1980]. With no applied field, a high-pass characteristic is obtained, while a

Fig. 7.14. Hologram recording configuration for BSO [Huignard, 1981].

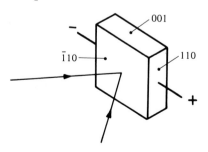

Fig. 7.15. Record-erase cycles in BSO at different applied fields. Sample size $10 \times 10 \times 10$ mm, $\lambda = 514$ nm [Huignard & Micheron, 1976].

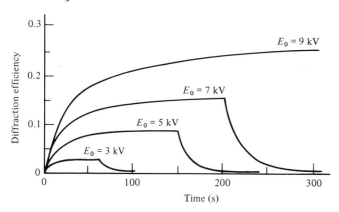

field of 600 V/mm gives a low-pass characteristic. An applied field of 200 V/mm results in a flat MTF.

Interesting optical effects are observed with these crystals since they are also strongly optically active [Miridonov, Petrov & Stepanov, 1978]. As a result, the transmitted beam, the diffracted beam and the scattered light exhibit differences in polarization. These can be exploited to minimize scatter noise, even in a nearly on-axis recording configuration, by the introduction of a sheet polarizer in the image plane [Herriau, Huignard & Aubourg, 1978].

The only problem with these materials is that optical readout is destructive. This is usually overcome by storing the reconstructed image on a vidicon memory tube for subsequent observation. The characteristics of these materials, which are the best currently available for read–write volume holographic storage, have opened up several interesting possibilities, including the continuous generation of a phase-conjugate wavefront (see section 12.5), real-time interferometry of vibrating objects (see section 15.1.2) and speckle reduction by real-time integration of a number of coherent images having independent speckle patterns [Huignard, Herriau, Pichon and Marrakchi, 1980].

8

Holograms for displays

An obvious application of holography is in displays. With the availability of suitable photographic emulsions coated on glass plates in sizes up to 1.5 × 1.0 m, several striking displays have been made using holograms. Some of the problems involved and techniques used to solve them have been described by Fournier, Tribillon & Viénot [1977] as well as by Bjelkhagen [1977a]. Even larger displays are possible with projection techniques using a reflecting lenticular screen (see, for example, Okoshi [1977]).

However, conventional holograms have several drawbacks [Benton, 1975]. Some of these are the limited angle over which the image can be viewed, low image luminance and the need to use a monochromatic source to illuminate the hologram, as well as the necessity to illuminate the subject with laser light when recording the hologram. This chapter discusses some of the techniques developed for holographic displays which have overcome these limitations (see also Benton [1980]).

8.1. 360° holograms

Holograms which give a 360° view of the object can be recorded using either four or more plates [Jeong, Rudolf & Luckett, 1966; Chau, 1970] or a cylinder of film to surround the object [Hioki & Suzuki, 1965; Jeong, 1967; Stirn, 1975; Upatnieks & Embach, 1980].

A very simple optical system for this purpose [Jeong, 1967] is shown in fig. 8.1. In this, the object is placed at the centre of a glass cylinder which has a strip of holographic film taped to its inner surface with the emulsion side facing inwards. The laser beam, which is expanded by a microscope objective and filtered by a pinhole, is incident on the object from above. The central portion of the beam illuminates the object, while the outer portions which fall directly on the film constitute the reference beam.

To view the reconstructed image, the processed film is put back in its original position in the recording setup after the object has been removed, and illuminated once again with the same laser.

A rather more efficient setup is obtained if the object is mounted on a plane or convex mirror. In this case the light reflected from the mirror constitutes the reference beam. However, the simpler optical system described has the advantage that it gives a very clean hologram, since it avoids noise due to defects in the mirror.

8.2. Double-sided holograms

While cylindrical holograms reconstruct images which can be viewed over a complete circle, they have the disadvantages that they are quite bulky and need a special illumination system. A more compact alternative which approximates to a full view of the image is a flat hologram which, when viewed from one side, displays the front of the object and, when viewed from the other side, displays its back [George, 1970]. If the positions of the two reconstructed images are properly chosen, the original spatial perspective is preserved, and the hologram appears to contain the object.

A typical sequence to produce a reflection hologram of this type is shown in fig. 8.2. In the first step, a transmission hologram is made of the object on plate 1 with the reference beam R1. Then, with the illuminated object still in position, plate 2 is exposed with the reference beam R2. After this, the object is removed and plate 1 is illuminated with the conjugate of R1, so that it forms a real image of the object in the same position. Another exposure is then made on plate 2 with another reference beam R3. Plate 2, after it is

Fig. 8.1. Setup for making a 360° hologram [Jeong, 1967].

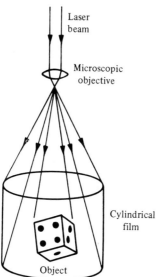

processed, is the final double-sided hologram. When it is illuminated, as shown in fig. 8.2(*d*), with two beams of white light, R2 and R4 (which is the conjugate of R3), it can be viewed from the two sides and shows the front and back of the object in their correct spatial relationship to each other.

This technique has been combined with the rainbow hologram technique (see section 8.5) by Hariharan [1977*b*], to produce double-sided transmission holograms which reconstruct bright images with a white-light source.

8.3.   Composite holograms

Another method which can be used to produce a flat hologram that reconstructs views of an object from a wide range of angles is to record a composite hologram [King, 1968].

Fig. 8.2. Steps involved in the production of a double-sided reflection hologram [George, 1970].

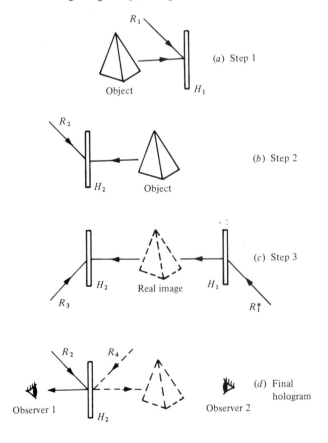

For this, a normal off-axis reference beam setup is used, but the object is placed on a turntable that can be rotated about a vertical axis and a mask containing a vertical slit is placed just in front of the photographic plate so that only a narrow strip of the plate is exposed. A series of holograms is then recorded on the plate. Between successive exposures, the mask is translated along the horizontal axis by the width of the slit, while the object is rotated through a small angle in the opposite direction. When the final composite hologram is illuminated, the observer sees a single virtual image of the object. However, if he moves his head from side to side, the reconstructed image appears to rotate about a vertical axis so that the effective angle of viewing can be as much as 360°.

A drawback of this arrangement is that there is inevitably some hyperstereoscopic distortion (an exaggerated effect of depth) when the image is viewed with both eyes. This can be minimized if the distance from which the image is viewed is increased. Alternatively, the composite hologram can be recorded with a horizontal slit which is moved vertically, so that both eyes view the image through a single hologram strip. In this case, of course, vertical parallax is lost.

8.4.   Holographic stereograms

It is also possible to synthesize a hologram that reconstructs an acceptable three-dimensional image from two-dimensional views of an object from different angles [McCrickerd & George, 1968; De Bitteto, 1969].

For this, a series of photographs of the subject is taken from equally spaced positions along a horizontal line. Contiguous, narrow, vertical strip holograms are then recorded of these photographs on a photographic plate, as shown in fig. 8.3. When this holographic stereogram is illuminated with a point source of monochromatic light, the viewer sees a three-dimensional image. The image lacks vertical parallax, but it exhibits horizontal parallax over the range of angles covered by the original photographs.

The obvious advantage of this technique over recording a hologram directly is that a laser is required only for the second step. White light can be used to illuminate the subject in the first step, so that holographic stereograms can be made of quite large scenes and even of living subjects.

Variations of this technique have been described by a number of authors. In one [Redman, Wolton & Shuttleworth, 1968], successive frames are stored as image holograms covering the whole photographic plate. Angular multiplexing is achieved by rotating the plate, along with the reference beam, between exposures by an angle corresponding to the original angular movement of the camera. Haig [1973] used a modified setup in which the reference beam is incident at an angle in the vertical plane, permitting it to move with the plate and allowing the use of a white-light source to illuminate the final image hologram. However, with these, the number of frames that can be multiplexed is limited by the drop in diffraction efficiency with multiple exposures.

Holographic stereograms have significant potential applications in three-dimensional displays of x-ray and ultrasonic images. Techniques for this purpose have been studied by Kasahara, Kimura, Hioki & Tanaka [1969], Groh & Kock [1970], Sopori & Chang [1971] and Kock & Tiemens [1973]. A simple arrangement using focused images which only requires translation of an aperture stop has been described by Prikryl & Kvapil [1980].

8.5.   The rainbow hologram

A major advance in the technology of holograms for displays was the development by Benton [1969, 1977] of a new type of transmission hologram capable of reconstructing a bright, sharp, monochromatic image when illuminated with white light. In this technique, part of the information content of the hologram is deliberately sacrificed to gain these advantages.

What is given up is parallax in the vertical plane; this is relatively unimportant, as pointed out earlier (see section 8.4), since depth perception depends essentially upon horizontal parallax. In return, a white-light source can be used for reconstruction, and there is a considerable gain in the brightness of the reconstructed image.

Fig. 8.3. Optical system used to record a holographic stereogram from a series of two-dimensional transparencies [De Bitteto, 1969].

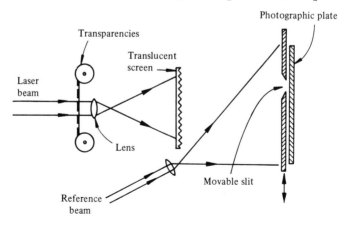

As shown schematically in fig. 8.4(a), the first step in making a rainbow hologram is to record a conventional transmission hologram of the object. This hologram is then illuminated, as shown in fig. 8.4(b), by the conjugate of the original reference wave. This generates a diffracted wave which is the conjugate of the original object wave and produces a real image of the object with unit magnification.

Fig. 8.4. Steps involved in the production of a rainbow hologram [Hariharan, 1983].

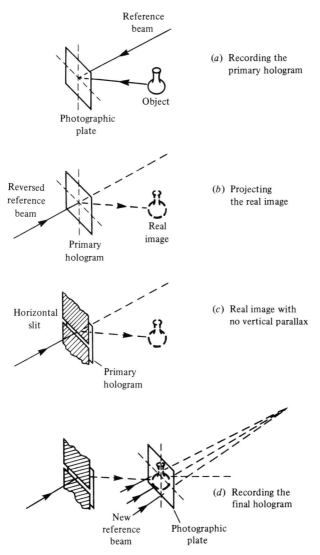

(a) Recording the primary hologram

(b) Projecting the real image

(c) Real image with no vertical parallax

(d) Recording the final hologram

Vertical parallax is then eliminated from this image, as shown in fig. 8.4(c), by a horizontal slit placed over the primary hologram. In effect, this limits the range of angles in the vertical plane from which the real image can be viewed, without restricting the range of viewing angles in the horizontal plane.

A second hologram is now recorded of this real image, as shown in fig. 8.4(d). The reference beam for this hologram is a convergent beam inclined in the vertical plane, and the hologram plate is placed so that the real image formed by the first hologram is very close to it.

When the final hologram is illuminated with the conjugate of the reference beam used to make it, it forms an orthoscopic real image of the object near the hologram, as shown in fig. 8.5(a). In addition, it also forms a real image of the slit placed across the primary hologram. All the light diffracted by the hologram passes through this slit pupil. Hence, if the observer's eyes move outside this slit pupil the image disappears. This corresponds to the almost complete elimination of vertical parallax.

Fig. 8.5. Reconstruction of the image by a rainbow hologram (a) with monochromatic light and (b) with white light [Hariharan, 1983].

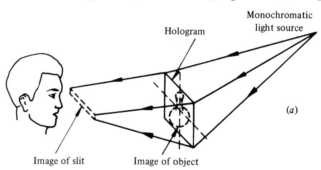

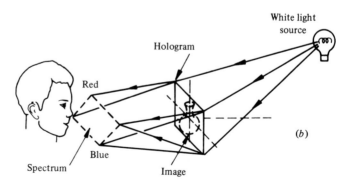

However, if the hologram is viewed from within the slit pupil a very bright image of the object is seen.

With a white-light source, the slit image is dispersed in the vertical plane, as shown in fig. 8.5(*b*), to form a continuous spectrum. An observer whose eyes are positioned at any part of this spectrum then sees a sharp three-dimensional image of the object in the corresponding colour.

Figure 8.6 shows an optical system which permits both steps of the

Fig. 8.6. Optical arrangement used to produce a rainbow hologram [Hariharan, 1977*b*].

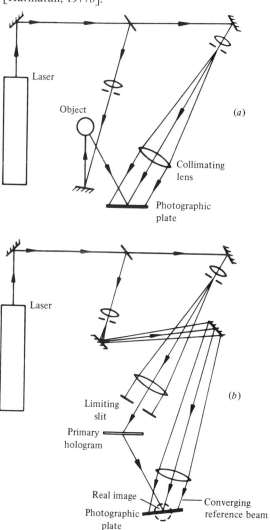

process to be carried out with a minimum of adjustments. A collimated reference beam is used, as shown in fig. 8.6(*a*), to record the primary hologram. When the setup is modified, as shown in fig. 8.6(*b*), for the second stage of the process, it is only necessary to turn the primary hologram through 180° about an axis normal to the plane of the figure and replace it in the plate holder. An undistorted real image is then projected into the space in front of the primary hologram. Vertical parallax is eliminated by a limiting slit, a few millimetres wide, placed over the collimating lens with its long dimension normal to the plane of the figure (this corresponds to the horizontal in the final viewing geometry). A convergent reference beam is used to record the final hologram, which, after processing, is reversed for viewing. When it is illuminated with a divergent beam from a point source of white light, an orthoscopic image of the object is formed, and a dispersed real image of the limiting slit is projected into the viewing space.

8.5.1. One-step rainbow holograms

A rainbow hologram can also be produced in a single step from real images of the object and the slit produced by an optical system [Benton, 1977; Chen & Yu, 1978]. An optical setup for this is shown in fig. 8.7. In this case a lens is used to form an orthoscopic image of the object at

Fig. 8.7. Optical system used to record a rainbow hologram in a single step [Yu, Tai & Chen, 1980].

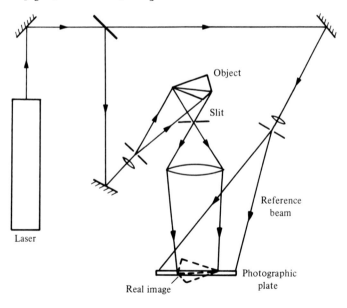

unit magnification just in front of the hologram plate. Unit magnification must be used if a three-dimensional image is to be free from distortion in depth, since the longitudinal magnification is the square of the transverse magnification (see section 3.2.3). A narrow slit is placed between the object and the focal plane of the lens, so that it is imaged into the viewing space at a suitable distance from the hologram. A diverging reference beam can be used in this case, corresponding to that used for reconstruction.

The only disadvantage of this technique is that the field of view is limited by the aperture of the lens. To obtain a reasonable field of view, an imaging lens with a large aperture and a relatively small focal length is required; this makes an optical system for the production of large holograms prohibitively expensive.

While methods have been suggested to minimize this problem [Tamura, 1978b; Benton, Mingace & Walter, 1979], a better alternative is to use a large concave spherical mirror [Hariharan, Hegedus & Steel, 1979] as described in section 9.5.

8.5.2. Image blur in the rainbow hologram

The image formed by a rainbow hologram is free from speckle, because it is reconstructed with an incoherent source, and usually appears quite sharp to the naked eye. However, there is always some image blur. The extent of this blur depends on the recording geometry and the size of the source used to illuminate the hologram and has been analysed by Wyant [1977] and by Tamura [1978a].

The primary cause of image blur is the finite wavelength spread in the image. To calculate this, consider a rainbow hologram made with the setup shown in fig. 8.8. If the angles made with the axis by the rays from the primary hologram to the final rainbow hologram are small compared to the

Fig. 8.8. Analysis of image blur in a rainbow hologram [Wyant, 1977].

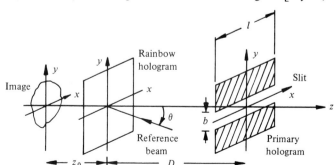

reference beam angle θ, the wavelength spread observed when the rainbow hologram is illuminated with white light is

$$\Delta\lambda = (\lambda/\sin\ \theta)[(b/D) + (a/D)], \tag{8.1}$$

where λ is the mean wavelength of the reconstructed image, b is the width of the slit, a is the diameter of the pupil of the eye, and D is the distance between the primary hologram and the final rainbow hologram. The two terms, (b/D) and (a/D) in (8.1) correspond to the angular subtense of the slit and the eye pupil measured from the hologram during recording and reconstruction, respectively. Typically, if $\lambda = 500$ nm, $\theta = 45°$, $D = 300$ mm and $a = b = 3$ mm, than $\Delta\lambda = 14$ nm.

The image blur $\Delta y_{\Delta\lambda}$ due to this wavelength spread is then given by the relation

$$\begin{aligned}\Delta y_{\Delta\lambda} &= z_0(\Delta\lambda/\lambda)\sin\theta, \\ &\approx z_0(a+b)/D, \end{aligned} \tag{8.2}$$

where z_0 is the distance of the image from the hologram. If the image is formed at a distance of 5 cm from a hologram with the parameters listed earlier, the image blur is approximately 1 mm. This corresponds to an angular blur of about 3 mrad at the eye pupil, which is acceptable.

Again, when a light source of finite size is used to illuminate the final hologram, the image exhibits an angular blur equal to the angular spread ψ_s of the source, as viewed from the hologram. The resultant image blur is

$$\Delta y_s = z_0\psi_s. \tag{8.3}$$

If this is not to exceed the blur due to the wavelength spread, the source should be chosen so that

$$\psi_s < (a+b)/D. \tag{8.4}$$

The final source of image blur is diffraction at the slit; this is noticeable only when the width of the slit is very small. Assuming that the slit is imaged on the eye pupil and its width $b \ll D$, the image blur due to diffraction is approximately

$$\Delta y_b = 2\lambda(z_0 + D)/b. \tag{8.5}$$

For the hologram parameters listed earlier, the image blur due to wavelength spread is greater than that due to diffraction, as long as the slit width b is more than 1 mm.

8.5.3. Rainbow holograms with reduced image blur

Image blur due to the finite spread of wavelengths in the image is mainly in the vertical plane and is a minimum when the image is formed in the hologram plane. A simple modification to the optical arrangement gives reduced image blur for images formed at an appreciable distance from the hologram [Leith & Chen, 1978].

For this, a cylindrical lens is placed over the slit aperture in front of the primary hologram, as shown in fig. 8.9, so that the image is brought to a focus in the vertical plane at the plate used to record the final hologram, without affecting the focus in the horizontal plane. While the reconstructed image has considerable astigmatism, this does not create any problems for the viewer. However, the image is quite sharp and free from colour blur. This technique does not, of course, increase the depth over which a sharp image can be obtained, but it permits a distant plane to be sharply imaged at the expense of a sacrifice in resolution for objects normally imaged in the plane of the hologram.

This technique has been extended to the one-step hologram process by Chen [1979] and Yu, Ruterbusch & Zhuang [1980]. Apart from permitting the use of a much wider slit, it also produces a higher object-beam intensity while recording the final rainbow hologram.

8.5.4.   Image luminance in the rainbow hologram

It has been shown [Hariharan, 1978] that with a recording system of the type shown in fig. 8.8, the average luminance of the image reconstructed at a wavelength λ by a rainbow hologram is given by the expression

$$L_v = [(D + z_0)/D][(A_H/A_I)(1/\psi)\Lambda\varepsilon E_\lambda K_\lambda], \tag{8.6}$$

where A_H is the area of the hologram that diffracts light, A_I is the area of the image, $\psi = l/(D + z_0)$ is the range of viewing angles in the horizontal plane, Λ is the average spacing of the fringes in the hologram, ε is the diffraction efficiency of the hologram, E_λ is the spectral irradiance at the hologram due

Fig. 8.9. Optical system used to produce a rainbow hologram with reduced image blur [Leith & Chen, 1978].

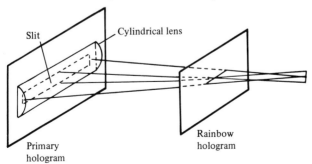

Slit

Cylindrical lens

Primary
hologram

Rainbow
hologram

to the beam illuminating it and K_λ is the spectral luminous efficacy of the radiation. If the image is close to the hologram, z_0 is negligible and $A_H \approx A_I$, since only the area of the hologram corresponding to the image diffracts light. The image luminance then becomes

$$L_v = \Lambda \varepsilon E_\lambda K_\lambda / \psi. \tag{8.7}$$

To maximize its luminance, the image should be located at the maximum distance from the hologram consistent with the permissible image blur, and the interbeam angle should be made as small as possible, consistent with the required field of view in the vertical plane.

8.6.   White-light holographic stereograms

An interesting development of these techniques has been the production by Cross of cylindrical holographic stereograms which, when illuminated with white light, reconstruct an almost monochromatic image (see Benton [1975]).

In the first step, the subject is placed on a slowly rotating turntable and a movie camera is used to make a record of a 120° or 360° rotation. Typically, three movie frames are exposed for each degree of rotation, so that the final movie sequence may contain up to 1080 frames.

The optical arrangement used to produce a holographic stereogram from this movie film is shown schematically in fig. 8.10. In this system, each frame is imaged in the vertical plane on to the hologram film. However, in the

Fig. 8.10. Optical arrangement used to produce a white light holographic stereogram [Huff & Fusek, 1980].

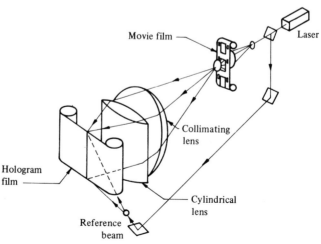

horizontal plane, the cylindrical lens brings all the rays leaving the projector to a line focus. A contiguous sequence of vertical strip holograms is then recorded of successive movie frames, covering the full range of views of the original subject, with a reference beam incident from below.

When the processed film is formed into a cylinder and illuminated with white light, a monochromatic image is seen which changes colour, as with any rainbow hologram, when the observer moves his head up or down. Due to the very large number of frames recorded, a modest amount of subject movement can be accommodated without destroying the stereoscopic image.

8.6.1. Image distortion in holographic stereograms

The images reconstructed by holographic stereograms can exhibit quite serious distortion if the recording and viewing geometry is not properly chosen. The anamorphic distortion observed with a plane holographic stereogram has been analysed by Glaser [1973] and Glaser & Friesem [1977], who have shown that it vanishes when

$$b(a-d)/a(b-d) = 1, \qquad (8.8)$$

where a, b and d are the distances from the viewing plane to the input transparency, the apparent location of the image, and the hologram respectively.

The case of a cylindrical holographic stereogram has been studied by Honda, Okada & Tsujiuchi [1981]. In this case the distortion is a minimum when

$$d = 2r/m, \qquad (8.9)$$

where d is the distance from the camera to the centre of rotation of the subject, r is the radius of the cylindrical hologram and m is the image magnification $(m < 1)$. Some optical techniques which can be used to compensate for distortion in the horizontal plane, yielding a wider field, have been outlined by Huff & Fusek [1981].

With a white-light stereogram, an additional cause of distortion is the fact that unless the hologram is viewed from a position corresponding to the focus of the field lens in the recording setup, different parts of the image are reconstructed in different colours. This distortion is particularly noticeable when the observer is too close to the hologram but is not so objectionable when he is further away [Okada, Honda & Tsujiuchi, 1981].

Okada, Honda & Tsujiuchi [1982] have also calculated the image blur in a white-light stereogram when a source of finite size is used to illuminate it. With a vertical line source, the image blur near the axis, as in a rainbow hologram, is influenced only by the width of the source; however, near the edges, the blur increases rapidly with the length of the source.

8.7.   Holographic movies

A moving three-dimensional image can be produced if a sequence of holograms of a scene is presented to an observer. Because the reconstructed image is not affected by movement of the hologram, provided an approximation to a Fourier transform setup is used (see sections 2.3 and 2.4), the film on which the holograms are recorded can be moved continuously.

In early attempts at holographic cinematography (see, for example, De Bitteto [1970]), the reconstructed image was viewed through the film by one person at a time. While this is acceptable for technical studies such as microscopic studies of marine plankton [Heflinger, Stewart & Booth, 1978], it is a serious limitation in the entertainment field. Work in the USSR [Komar, 1977] has therefore been based on a projection technique. In this, a lens with a diameter of 200 mm is used to record a series of image holograms of the scene on 70 mm film. After processing, the reconstructed image is projected with an identical lens on to a special holographic screen which is equivalent to several superposed concave mirrors. Each of these holographic concave mirrors then forms a real image of the projection lens in front of a spectator, so that when he looks through this pupil he sees a full-size three-dimensional image.

9

Colour holography

The fact that a multicolour image can be produced by a hologram recorded with three suitably chosen wavelengths was first pointed out by Leith & Upatnieks [1964].

The resulting hologram can be considered as made up of three incoherently superposed holograms. When it is illuminated once again with the three wavelengths used to make it, each of them is diffracted by the hologram recorded with it to give a reconstructed image in the corresponding colour. The superposition of these three images results in a multicolour image.

However, while multicolour holography was demonstrated at quite an early stage, several practical problems were encountered. These problems as well as recent advances which have made multicolour holography practical are described in this chapter (see also a review by Hariharan [1983]).

9.1. Light sources for colour holography

The most commonly used lasers for colour holography are the He–Ne laser ($\lambda = 633$ nm) and the Ar^+ laser which has two strong output lines ($\lambda = 514$ nm and 488 nm; see Table 5.1). The range of colours that can be reconstructed with these three wavelengths as primaries can be determined by means of the C.I.E. chromaticity diagram [The Optical Society of America, 1953] as shown in fig. 9.1. In this, points representing monochromatic light of different wavelengths constitute the horseshoe-shaped curve known as the spectrum locus; all other colours lie within this boundary. New colours obtained by mixing light of two wavelengths, such as 633 nm and 514 nm, lie on the straight line AB joining these primaries. When light with a wavelength of 488 nm is also used, any colour within the triangle ABC can be obtained.

A wider range of hues can be obtained if other laser lines are used, permitting a better choice of primaries. One of these is the He–Cd laser line ($\lambda = 442$ nm) which is a very attractive blue primary, but involves the use of one more laser. A simpler alternative is the blue line of the Ar^+ laser at a wavelength of 477 nm, which, at some sacrifice of power, can give a significant improvement in colour rendering as shown by the broken lines in fig. 9.1.

The Kr$^+$ laser line at 647 nm is a useful alternative to the red line of the He–Ne laser for recording large holograms, since the power available is much higher, and single-mode output can be obtained with an etalon.

Fig. 9.1. C.I.E. chromaticity diagram. The triangle ABC shows the range of hues that can be produced by a hologram illuminated with primary wavelengths of 633 nm, 514 nm and 488 nm, while the broken lines show the extended range possible if the blue primary is replaced by one at 477 nm. The chain lines enclose the range of hues which can be reproduced by a typical colour-television display [Hariharan, 1983].

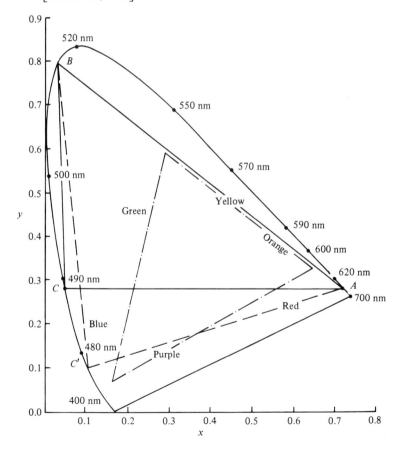

9.2. The cross-talk problem

A problem in multicolour holography is that each hologram diffracts not only light of the wavelength used to record it, but the other two wavelengths as well. As a result, a total of nine primary images and nine conjugate images are produced. Three of these give rise to a full-colour reconstructed image at the position originally occupied by the object. The remaining images resulting from light of one wavelength diffracted by a component hologram recorded with another wavelength are formed in other positions and overlap with and degrade the multicoloured image.

Several methods have been tried to eliminate these cross-talk images, including frequency multiplexing [Mandel, 1965; Marom, 1967], spatial multiplexing or coded reference beams [Collier & Pennington, 1967], and division of the aperture field [Lessard, Som & Boivin, 1973]. However, all these methods suffer from drawbacks such as a restricted image field, a reduction in resolution or a decrease in the signal-to-noise ratio. In addition, they need multiple laser wavelengths (or equivalent monochromatic light sources) to illuminate the hologram.

9.3. Volume holograms

The first methods to eliminate cross-talk which did not involve such penalties were based on the use of volume holograms.

A hologram recorded with several wavelengths in a thick medium contains a set of regularly spaced fringe planes for each wavelength. When this hologram is illuminated once again with the original multiwavelength reference beam, each wavelength is diffracted with maximum efficiency by the set of fringe planes created originally by it, producing a multicoloured image. However, the cross-talk images are severely attenuated since they do not satisfy the Bragg condition.

This principle was first applied by Pennington & Lin [1965] to produce a two-colour hologram of a transparency, and subsequently extended by Friesem & Fedorowicz [1966, 1967] to three-colour imaging of diffusely reflecting objects. The optical setup used by them is shown schematically in fig. 9.2. In it, blue and green light ($\lambda = 488$ nm and 514 nm) from an Ar$^+$ laser was mixed with red light ($\lambda = 633$ nm) from a He–Ne laser to produce two beams containing light of all three wavelengths. One beam was used to illuminate the object while the other was used as a reference beam, and the resulting hologram was recorded in a thick photographic emulsion. When this hologram was illuminated once again with a similar multicolour beam at the appropriate angle, a multicolour reconstructed image with negligible cross-talk was obtained.

9.4.   Volume reflection holograms

As we have seen earlier in Chapter 4, the highest wavelength selectivity is obtained with volume reflection holograms. Such holograms can even reconstruct a monochromatic image when illuminated with white light. Their use for multicolour imaging followed directly [Lin, Pennington, Stroke & Labeyrie, 1966; Upatnieks, Marks & Fedorowicz, 1966; Stroke & Zech, 1966].

For this, a volume reflection hologram is recorded with three wavelengths, so that one set of fringe planes is produced for each wavelength. When such a hologram is illuminated with white light, each set of fringe planes, because of its high wavelength selectivity, diffracts a narrow band of wavelengths centred on the original laser wavelength used to record it, giving a multicolour reconstructed image free from cross-talk.

However, early reflection holograms recorded on photographic emulsions and processed in the conventional manner had several drawbacks. The most serious of these was their low diffraction efficiency, which was aggravated by the fact that when more than one hologram is recorded in the same emulsion layer, the available dynamic range is shared between the recordings. As a result, the diffraction efficiency of each recording is reduced by a factor approximately equal to the square of the number of holograms [Collier, Burckhardt & Lin, 1971] (see section 4.10).

Another problem was the reduction in the thickness of a photographic emulsion layer which occurs during processing, due to removal of the

Fig. 9.2. Setup used to record a multicolour hologram of a diffusely reflecting object in a thick recording medium [Friesem & Fedorowicz, 1966].

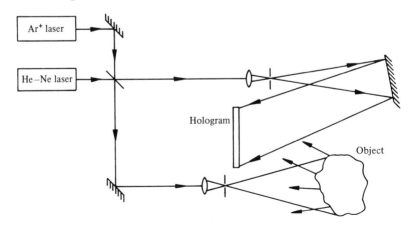

unexposed silver halide. This results in a reduction in the fringe spacing and a shift in the colour of the reconstructed image towards shorter wavelengths. Typically, a hologram recorded with red light ($\lambda = 633$ nm) reconstructs a green image ($\lambda \approx 530$ nm). While this shrinkage can be eliminated by soaking the emulsion in an aqueous solution of triethanolamine, $(CH_2OHCH_2)_3N$, before drying [Lin & Lo Bianco, 1967], this is not entirely satisfactory.

Recent work has shown better ways to overcome these problems. One approach has been the use of other recording materials such as dichromated gelatin sensitized with methylene blue [Kubota & Ose, 1979b]. Since multicolour reflection holograms recorded on this material have peak diffraction efficiencies as high as 40 per cent, they show great promise.

However, because of the relatively low sensitivity of other materials, photographic emulsions still remain an attractive recording medium for volume reflection holograms, and improved techniques have now made it possible to produce much brighter images with them. Thus, low-noise bleaches can be used to produce reflection holograms with improved diffraction efficiency [Hariharan, 1972; Phillips *et al.*, 1980]. Moreover, the use of relatively thin emulsion layers ($\approx 6 \mu m$ thick) which diffract a wider spectral bandwidth can give as high an image luminance as a thicker layer, with the advantage of lower scattering [Hariharan, 1972, 1979b].

Triethanolamine cannot be used to correct the shrinkage in thickness of such bleached holograms, since they then darken rapidly when exposed to light due to the formation of photolytic silver. A solution (≈ 10 per cent) of $D(-)$sorbitol$[CH_2OH(CHOH)_4CH_2OH]$ can be used instead [Hariharan, 1980a]; alternatively, a tanning developer can be used to minimize, or even eliminate, this shrinkage [Joly & Vanhorebeek, 1980].

Since such bleached reflection holograms are completely transparent at wavelengths outside the relatively narrow band which is diffracted, it is also possible to record the three component holograms for different primary wavelengths on two separate plates and superimpose them to make up the final multicolour hologram. This permits the use of different types of plates to record the component holograms, one with optimum characteristics for the red, and the other with optimum characteristics for the green and the blue. In addition, an improvement in image luminance by a factor of 2 or more is obtained if the three component holograms are divided between two plates in this manner, instead of being recorded in the same emulsion layer.

A further improvement in image luminance can be obtained if the diffracted light from the hologram is concentrated into a smaller solid angle. For this, a hologram is recorded, not of the original object, but of a real image of the object projected either by another hologram or by an optical system whose effective aperture is limited by a suitably shaped stop as shown in fig. 9.3. This is the same principle which has been exploited in the rainbow hologram. The gain in image luminance is proportional to the reciprocal of the available solid angle of viewing [Hariharan, 1978]. Normally, a gain in image luminance by a factor of 3 or 4 can be obtained, without any loss in convenience, if the range of viewing angles in the vertical plane is restricted to about 15°.

Typically, the red component hologram is recorded with a He–Ne laser ($\lambda = 633$ nm) and 8E75 plates, while an Ar^+ laser is used with 8E56 plates for the green ($\lambda = 514$ nm) and blue ($\lambda = 488$ nm) exposures. As shown in fig. 9.4, the 8E75 plate is exposed with the emulsion side towards the reference beam, while the 8E56 plate is exposed with the emulsion side facing the

Fig. 9.3. Optical system used for recording multicolour reflection holograms with increased image luminance [Hariharan, 1980a].

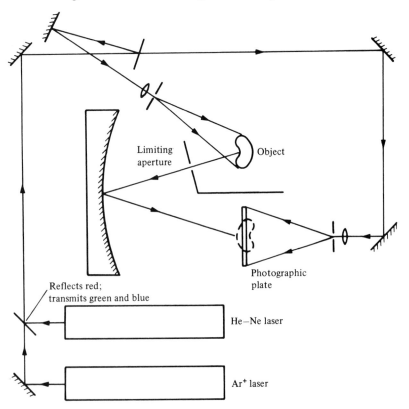

Limiting aperture

Object

Photographic plate

Reflects red; transmits green and blue

He–Ne laser

Ar^+ laser

mirror. To compensate for the thickness of the plates, the plate holder is mounted on a micrometer slide and moved normal to its plane through a distance equal to $d[1-(1/n)]$, where d is the thickness of the plates and n is the refractive index of the glass, between the two sets of exposures.

After drying, the plates are assembled with the emulsion layers in contact and the reconstructed images are viewed, with the 8E56 plate in front and the 8E75 plate behind. This helps to equalize the diffraction efficiencies of the holograms.

9.5.   Multicolour rainbow holograms

A completely different approach was opened up by the extension of the rainbow hologram technique to three-colour recording [Hariharan, Steel & Hegedus, 1977; Tamura, 1977; Suzuki, Saito & Matsuoka, 1978].

Fig. 9.4. Schematic showing how a multicolour reflection hologram is built up from exposures on two plates [Hariharan, 1980*a*].

(*a*) Red image

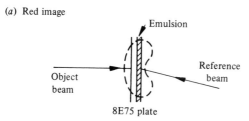

(*b*) Green/blue image

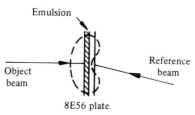

(*c*) Final hologram

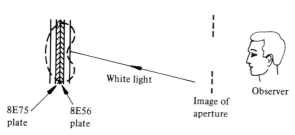

This made it possible to produce holograms that reconstruct very bright multicolour images when illuminated with a white-light source.

For this, three primary holograms (colour separations) are made from the object with red, green and blue laser light using an optical system similar to that shown in fig. 8.6. In the second stage, these primary holograms are used with the same laser sources to make a single hologram consisting of three superposed recordings. When this multiplexed hologram is illuminated with a white-light source, it reconstructs three superimposed images of the object. In addition, three spectra are formed in the viewing space. These are, as before, dispersed real images of the limiting slit. However, these spectra are displaced vertically with respect to each other, as shown in fig. 9.5 so that, in effect, each component hologram reconstructs an image of the limiting slit in its original position in the colour with which it was made. Accordingly, an observer viewing the hologram from the point where the spectra overlap sees three superimposed images of the object reconstructed in the colours with which the primary holograms were made.

Multicolour rainbow holograms can also be produced in a single step [Chen, Tai and Yu, 1978; Hariharan, Hegedus and Steel, 1979]. A typical optical system for this using a concave mirror is shown in fig. 9.6. The mirror used had a diameter of 600 mm and a radius of curvature of 550 mm, giving an angle of view of approximately 50°. The object, turned sideways, was placed on one side of the axis of the mirror so that its image was formed on the other side, at the same distance from the mirror. A vertical slit was placed between the object and the mirror, at such a distance from the mirror that a magnified image of it was formed in the viewing space at a convenient distance (≈ 1 m) from the hologram. To make the final hologram, three

Fig. 9.5. Reconstruction of a multicolour image by superimposed rainbow holograms [Hariharan, 1983].

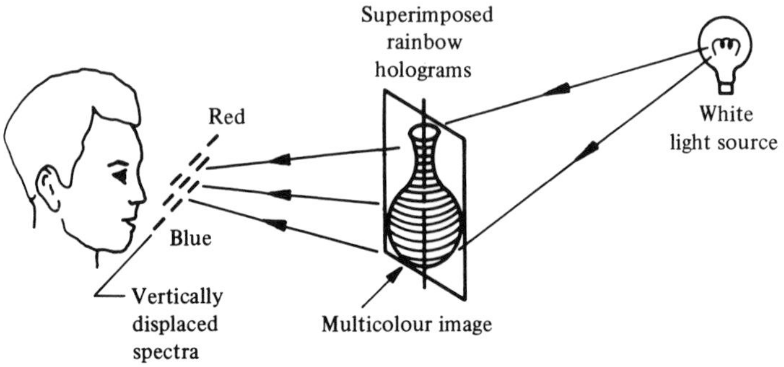

successive exposures were made, using red light ($\lambda = 633$ nm) from a He–Ne laser and green and blue light ($\lambda = 514$ nm and 488 nm) from an argon laser.

While, in principle, the three superimposed holograms making up the final multicolour rainbow hologram can be recorded on a single plate, there are several advantages in using a sandwich technique [Hariharan, Steel & Hegedus, 1977]. Besides making it possible to use different types of photographic plates whose characteristics are optimized for the different wavelengths, it also makes it much easier to match the diffraction efficiencies of the three individual holograms. In addition, it also gives much brighter images, since, with bleached holograms, the loss in diffraction efficiency due to multiplexing three holograms on a single plate can be partially avoided.

The simplest method is to use two plates. In this case, as shown in fig. 9.7(*a*) on the left, the red component hologram is recorded on a red-sensitive plate (8E75), which is loaded into the plate holder with the emulsion facing

Fig. 9.6. Layout of the optical system used to produce multicolour rainbow holograms in a single step with a concave mirror [Hariharan, Hegedus & Steel, 1979].

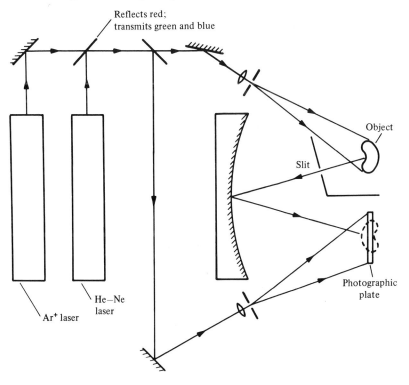

Reflects red;
transmits green and blue

Object

Slit

Photographic plate

He–Ne
laser

Ar$^+$ laser

forward and with a clear glass plate of the same thickness in front of it. After this, the blue and green component holograms are recorded on an orthochromatic plate (8E56) without any antihalation coating. This is loaded into the plate holder as shown in fig. 9.7(*a*) on the right, with the emulsion side facing backward and with a clear glass plate of the same thickness behind it. Finally, as shown in fig. 9.7(*b*), the two processed plates are cemented together with their emulsions in contact to form the final multicolour hologram. Registration of the two plates need be done only to an accuracy determined by the residual image blur and is automatic if the plate holder is used as an assembly jig. This technique gives an improvement in diffraction efficiency by a factor of 2, or more, over that possible if the two holograms are recorded in the same emulsion.

However, the best results are obtained if the three recordings are made on separate plates and superimposed; in this case an improvement in diffraction efficiency by a factor of 3 is possible. However, to compensate for the thickness of the glass plates and ensure that the reconstructed images coincide in depth, the plateholder must be moved forward or backward through a distance (d/n), where d is the thickness of the plates and n is the refractive index of the glass, between exposures.

Multicolour rainbow holograms give bright images even when illumi-

Fig. 9.7. Schematic of the sandwich technique used to make multicolour holograms [Hariharan, Steel & Hegedus, 1977].

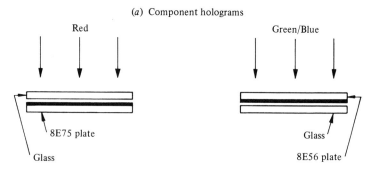

(*a*) Component holograms

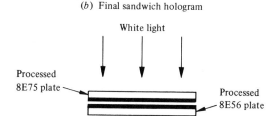

(*b*) Final sandwich hologram

nated with an ordinary tungsten lamp. As can be seen from fig. 9.8, the images exhibit high colour saturation and are free from cross-talk. Problems with emulsion shrinkage are eliminated, since volume effects are not involved, and shrinkage primarily affects only the thickness of the emulsion. As with any rainbow hologram, the colours of the image change with the viewing angle in the vertical plane. This change can be utilized effectively in some types of displays, but, where necessary, it can be kept within acceptable limits by optimization of the length of the spectra projected into the viewing space and the use of baffles to define the available range of viewing angles in the vertical plane.

Fig. 9.8. 'Cubes'. Image reconstructed by a multicolour rainbow hologram.

9.6.   Pseudocolour images

The fact that the colour information in a hologram is recorded only in the form of specific carrier fringe frequencies suggests the possibility of using a single laser wavelength and generating the different carrier frequencies by some other means.

One method is to record three superimposed rainbow holograms, changing the angle of the reference beam between exposures [Tamura, 1978*a*]. Alternatively, the position of the limiting slit can be changed between exposures [Vlasov, Ryabova & Semenov, 1977; Yan-Song, Yu-Tang & Bi-Zhen, 1978].

A problem with pseudocolour rainbow holograms is that the images reconstructed in a different colour from that used to record the hologram are displaced with respect to an image of the same colour. The displacement in the vertical plane is

$$\Delta y = z_0 (2\Delta\lambda/\lambda) \tan^3 \theta, \tag{9.1}$$

and the longitudinal displacement is

$$\Delta z = z_0 (2\Delta\lambda/\lambda), \tag{9.2}$$

where z_0 is the distance of the image from the hologram, λ is the wavelength with which the hologram is recorded and $(\lambda + \Delta\lambda)$ is the wavelength at which the image is reconstructed. These displacements can be brought within acceptable limits if the image is formed sufficiently close to the hologram.

With volume reflection holograms, changing the angle between the reference and object beams has little effect on the colour of the reconstructed image. However, its colour is affected by changes in the thickness of the recording medium and these changes can be controlled and used to produce pseudocolour images [Hariharan, 1980*c*].

If a He–Ne laser is used, the red component hologram is recorded first on a plate exposed with the emulsion side towards the reference beam. This plate is bleached and processed to eliminate emulsion shrinkage. The green and blue component holograms are then recorded on another plate exposed with the emulsion side towards the object beam. The green component hologram is exposed first with the emulsion in its normal condition. After this the emulsion is soaked in a 3 per cent solution of triethanolamine and dried in darkness. The blue component hologram is then exposed on the swollen emulsion. Normal bleach processing then eliminates the triethanolamine and produces the usual shrinkage. Accordingly, the first exposure yields a green image, while the second produces one at an even shorter wavelength, that is to say, a blue image.

After drying, the plates are cemented together with the emulsion layers in contact. The images are viewed with the hologram reconstructing the green and blue images in front and the hologram reconstructing the red image behind.

In this case also, the images reconstructed at wavelengths differing from that used to record the holograms undergo shifts which depend on μ, the ratio of these wavelengths [Hariharan, 1976a, 1980c]. To avoid lateral misregistration, the lateral magnification M_{lat} must be independent of μ; this is possible if the hologram is illuminated with a parallel beam. While it is not possible to eliminate longitudinal misregistration and longitudinal distortion simultaneously, acceptable results can be obtained if the image is centred on the hologram plane.

9.7.   Achromatic images

Holograms that reconstruct an almost white image when illuminated with white light have the advantage that they can produce very bright images, since they use the entire output of the source.

As pointed out in section 3.5.3, a very nearly achromatic image of an object of limited depth can be produced by an image hologram. However, to produce an achromatic image of an object with appreciable depth, some technique of dispersion compensation must be used.

One of the earliest methods [De Bitteto, 1966] used a plane diffraction grating with a line spacing equal to the average fringe spacing in the hologram, as shown in fig. 9.9, to provide equal but opposite angular dispersion. To avoid directly transmitted light spilling into the field of view, a light shield consisting of a set of parallel baffles, rather like a venetian

Fig. 9.9. Production of an achromatic image by dispersion compensation using a diffraction grating [De Bitteto, 1966].

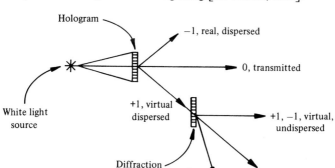

blind, can be used between the hologram and the grating [Burckhardt, 1966*b*].

Brighter reconstructed images can be obtained with techniques using rainbow holograms. The simplest method is merely to use a vertical line source to illuminate the hologram [Benton, 1969, 1977]. This results in a series of overlapping spectra projected into the viewing space, so that a near-white image is obtained.

However, there is still some residual colour blur because the images reconstructed by different wavelengths have different magnifications and are formed at different distances from the hologram. This is also true of the slit images constituting the spectrum projected into the viewing space. To produce a truly achromatic image, the red, green and blue images must coincide; this means that the overlapping spectra must lie along the same line.

To achieve this result, Benton [1978] has used a multiply exposed holographic lens, which produces the effect of a series of point sources of light located at suitable angles and distances, in combination with a narrow strip of the first hologram to make a second hologram. This hologram

Fig. 9.10. Head of Aphrodite, 'The Bartlett Head' (Greek, *ca.* 325–300 B.C.: the Boston Museum of Fine Arts). Achromatic white-light transmission hologram 270 mm × 320 mm produced by S. A. Benton, J. L. Benton, H. S. Mingace Jr., and W. Walter at the Research Laboratories of the Polaroid Corporation, Cambridge, USA, 1977.

Fig. 9.11. Setup using a one-dimensional diffuser to produce a hologram that reconstructs an achromatic image [Leith, Chen & Roth, 1978].

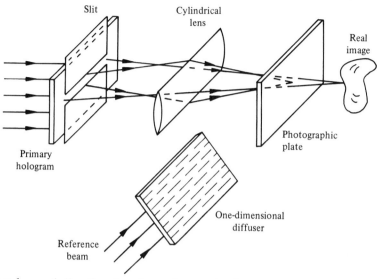

cannot be used directly, since it produces a badly blurred image, but it is illuminated with monochromatic light and a third hologram is made of the image reconstructed by it. When this final hologram is illuminated with white light, it reconstructs a set of overlapping spectra in the viewing space which coincide at appropriate wavelengths, so that an achromatic image is obtained over a wide range of viewing angles.

The image reconstructed by an achromatic hologram produced with this technique is shown in fig. 9.10.

Another technique described by Leith, Chen & Roth [1978] uses a one-dimensional diffuser to generate multiple reference beams. As shown in fig. 9.11, the primary hologram is masked by a horizontal slit and forms a real image at some distance from the hologram plate. However, the introduction of a cylindrical lens causes this image to be focused in the vertical plane at the hologram itself. The phase plate in the reference beam diffuses it in the vertical direction and transmits without scatter in the horizontal direction. As a result, light of any colour incident on the hologram is scattered through a range of angles in the vertical plane determined by the extent of the diffuser. This has little effect on the image resolution, since the image is focused in the vertical plane at the hologram, but it washes out the colour effects normally obtained with a rainbow hologram.

10

Computer-generated holograms

Holograms produced by means of a computer permit the generation of wavefronts with any prescribed amplitude and phase distribution; they are therefore extremely useful for testing optical surfaces as well as in applications such as laser-beam scanning and optical spatial-filtering (see sections 12.10 and 13.2).

The production of holograms using a digital computer has been discussed in detail by Lee [1978], Yaroslavskii & Merzlyakov [1980] and Dallas [1980]. Basically, it involves two major steps.

The first is to calculate the complex amplitude of the object wave at the hologram plane; for convenience this is usually taken to be the Fourier transform of the complex amplitude in the object plane. It can be shown, by means of the sampling theorem, that if the object wave is sampled at a sufficiently large number of points (see Appendix A.1), this can be done with no loss of information. Thus, if an image consisting of $N \times N$ resolvable elements is to be reconstructed, the object wave is sampled at $N \times N$ equally spaced points, and the $N \times N$ complex coefficients of its discrete Fourier transform are evaluated. This can be done quite easily with a computer program using the fast Fourier transform algorithm [Cochran *et al.*, 1967] for arrays containing as many as 1024×1024 points.

The second step involves using the computed values of the discrete Fourier transform to produce a transparency (the hologram) which reconstructs the object wave when it is suitably illuminated.

Two approaches have been followed for this purpose. The first is analogous to off-axis optical holography (see section 2.2). The complex amplitudes of a plane reference wave and the object wave in the hologram plane are added and the squared modulus of their sum is evaluated. This permits the production of a transparency whose amplitude transmittance is real and positive everywhere. A variation [Burch, 1967] makes use of the fact that in the expression on the right hand side of (2.7) the first two terms do not contribute to the reconstructed image. Hence, it is only necessary to compute the last term at regular intervals and add a constant bias to all the samples to make them positive.

An alternative approach, which is possible only with a computer-generated hologram, is to produce a transparency which records both the amplitude and the phase of the object wave in the hologram plane. This can be thought of as the superposition of two transparencies, one of constant thickness having a transmittance proportional to the amplitude of the object wave, and the other having thickness variations corresponding to the phase of the object wave, but no transmittance variations. Such a hologram has the advantage that it forms a single image, on-axis.

In either case, the computer is used to control a plotter which produces a large-scale version of the hologram. This is photographically reduced to produce the required transparency. An optical system similar to that shown in fig. 2.7 is then used to reconstruct the object wavefront.

10.1. The binary detour-phase hologram

While it is possible to use an output device with gray scale capabilities to produce the hologram, a considerable simplification results if the amplitude transmittance of the hologram has only two levels – either zero or one. Such a hologram is called a binary hologram.

The best known hologram of this type is the binary detour-phase hologram [Brown & Lohmann, 1966, 1969] which is made without explicit use of a reference wave or bias. To produce such a hologram, the output format covered by the plotter is divided into $N \times N$ cells, which correspond to the $N \times N$ coefficients of the discrete Fourier transform of the complex amplitude in the object plane. Each complex Fourier coefficient is then represented by a single transparent area within the corresponding cell, whose size is determined by the modulus of the Fourier coefficient, while its position within the cell represents the phase of the Fourier coefficient. This method derives its name from the fact that a shift of the transparent area in each cell results in the light transmitted by it travelling by a longer or shorter path to the reconstructed image. Figure 10.1(*a*) shows a typical binary detour-phase hologram of the letters ICO, while fig. 10.1(*b*) shows the image produced by it. The first-order images are those above and below the central spot; in addition, higher-order images are seen due to nonlinear effects (see section 10.4).

To understand how this method of encoding the phase works, consider a rectangular opening ($a \times b$) in an opaque sheet (the hologram) centred on the origin of coordinates, as shown in fig. 10.2, and illuminated with a uniform coherent beam of light of unit amplitude. The complex amplitude $U(x_i, y_i)$ at a point (x_i, y_i) in the diffraction pattern formed in the far field is given by the Fourier transform of the transmitted amplitude and is

$$U(x_i, y_i) = ab \operatorname{sinc}(ax_i/\lambda z) \operatorname{sinc}(by_i/\lambda z), \qquad (10.1)$$

where $\operatorname{sinc} x = (\sin \pi x)/\pi x$.

We now assume that the centre of the rectangular opening is shifted to a point $(\Delta x_0, \Delta y_0)$ and the sheet is illuminated by a plane wave incident at an angle. If the complex amplitude of the incident wave at the sheet is $\exp[i(\alpha \Delta x_0 + \beta \Delta y_0)]$, the complex amplitude in the diffraction pattern

Fig. 10.1. Binary detour-phase hologram: (*a*) the hologram; (*b*) the reconstructed image [Lohmann & Paris, 1967].

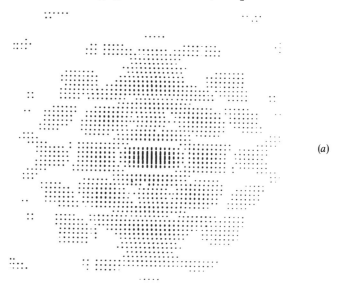

(*a*)

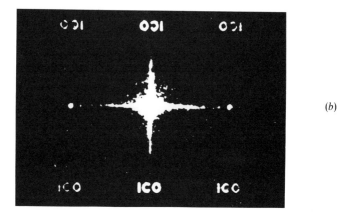

(*b*)

becomes

$U(x_i, y_i) = ab$ sinc $(ax_i/\lambda z)$ sinc $(by_i/\lambda z)$

$$\times \exp\left[i\left(\alpha + \frac{2\pi x_i}{\lambda z}\right)\Delta x_0 + i\left(\beta + \frac{2\pi y_i}{\lambda z}\right)\Delta y_0 \right],$$

$= ab$ sinc $(ax_i/\lambda z)$ sinc $(by_i/\lambda z)$

$$\times \exp[i(\alpha\Delta x_0 + \beta\Delta y_0)] \exp\left[i\left(\frac{2\pi}{\lambda z} x_i\Delta x_0 + \frac{2\pi}{\lambda z} y_i\Delta y_0\right)\right]. \quad (10.2)$$

If $ax_i \ll \lambda z$, $by_i \ll \lambda z$, (10.2) reduces to

$$U(x_i, y_i) = ab \exp[i(\alpha\Delta x_0 + \beta\Delta y_0)]\exp\left[i\left(\frac{2\pi}{\lambda z}x_i\Delta x_0 + \frac{2\pi}{\lambda z}y_i\Delta y_0\right)\right]. \quad (10.3)$$

If then, the computed complex amplitude of the object wave at a point $(n\Delta x_0, m\Delta y_0)$ in the hologram plane is

$$o(n\Delta x_0, m\Delta y_0) = |o(n\Delta x_0, m\Delta y_0)| \exp\left[i\phi(n\Delta x_0, m\Delta y_0)\right], \quad (10.4)$$

its modulus and phase at this point can be encoded, as shown in fig. 10.3, by making the area of the opening located in this cell equal to the modulus so that

$$ab = |o(n\Delta x_0, m\Delta y_0)|, \quad (10.5)$$

and displacing the centre of the opening from the centre of the cell by an amount δx_{nm} given by the relation

$$\delta x_{nm} = (\Delta x_0/2\pi)\phi(n\Delta x_0, m\Delta y_0). \quad (10.6)$$

To show the validity of this method of encoding, we consider the complex amplitude in the far field due to this opening, which is, from (10.3),

$$U_{nm}(x_i, y_i) = |o(n\Delta x_0, m\Delta y_0)|\exp[i\alpha(n\Delta x_0 + \delta x_{nm}) + i\beta m\Delta y_0]$$
$$\times \exp[(i2\pi/\lambda z)(nx_i\Delta x_0 + my_i\Delta y_0 + \delta x_{nm})]. \quad (10.7)$$

The total diffracted amplitude in the far field, which is obtained by

Fig. 10.2. Diffraction at a rectangular aperture.

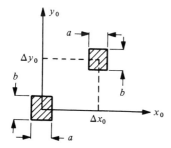

summing the complex amplitudes due to all the $N \times N$ openings, is therefore

$$
\begin{aligned}
U(x_i, y_i) = \sum_{n=1}^{N} \sum_{m=1}^{N} &|o(n\Delta x_0, m\Delta y_0)| \exp(i\alpha \delta x_{nm}) \\
&\times \exp[i(\alpha n \Delta x_0 + \beta m \Delta y_0)] \\
&\times \exp[(i2\pi/\lambda z)(nx_i \Delta x_0 + my_i \Delta y_0)] \exp[(i2\pi/\lambda z)\delta x_{nm}].
\end{aligned}
$$

(10.8)

If the dimensions of the cells and the angle of illumination are chosen so that

$$\alpha \Delta x_0 = 2\pi, \tag{10.9}$$

$$\beta \Delta y_0 = 2\pi, \tag{10.10}$$

and $$\delta x_{nm} \ll \lambda z, \tag{10.11}$$

(10.8) reduces to

$$
\begin{aligned}
U(x_i, y_i) = \sum_{n=1}^{N} \sum_{m=1}^{N} &|o(n\Delta x_0 + m\Delta y_0)| \exp[i\phi(n\Delta x_0, m\Delta y_0)] \\
&\times \exp[(i2\pi/\lambda z)(nx_i \Delta x_0 + my_i \Delta y_0)].
\end{aligned}
$$

(10.12)

This is the discrete Fourier transform of the computed complex amplitude in the hologram plane or, in other words, the desired reconstructed image.

Binary detour-phase holograms have several attractive features. It is

Fig. 10.3. Typical cell in a binary detour-phase hologram.

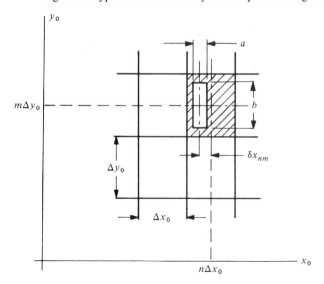

possible to use a simple pen-and-ink plotter to prepare the binary master, and problems of linearity do not arise in the photographic reduction process. Their chief disadvantage is that they are very wasteful of plotter resolution, since the number of addressable plotter points in each cell must be large to minimize the noise due to quantization of the modulus and the phase of the Fourier coefficients. When the number of quantization levels used is fairly large, this noise is effectively spread over the whole image field, independent of the form of the signal. However, when the number of phase-quantization levels is small, the noise terms become shifted and self-convolved versions of the signal, which are much more annoying [Goodman & Silvestri, 1970; Dallas, 1971a, b].

10.2. Generalized binary detour-phase holograms

In this method, as shown in fig. 10.4, rather than producing a single transparent area with variable size and position in the cell, corresponding to each Fourier coefficient, a combination of $p \times q$ transparent and opaque subcells is used [Haskell & Culver, 1972; Haskell, 1973]. This method permits finer quantization of both amplitude and phase, resulting in less noisy images. However, it is necessary for the computer to identify the proper binary pattern out of the 2^{pq} possible patterns, that is the best approximation to the desired complex Fourier coefficient, before plotting it.

10.3. Lee's method

An alternative method of recording a computer-generated hologram is due to Lee [1970]. In this, each cell on the hologram is divided into four equal-sized subcells arranged side by side, as shown in fig. 10.5(a). The four subcells then contribute phasor components with relative phases of 0°, 90°, 180° and 270°, because of their positions within the cell, the amplitude

Fig. 10.4. Typical cell in a generalized binary detour-phase hologram.

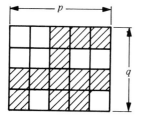

of each contribution being determined by the transmittance of that subcell. In actual use, two of the subcells in any cell are totally opaque, while the other two are partially transmitting. This gives complete control of both the amplitude and the phase of the resultant complex amplitude.

A simplification of this technique is due to Burckhardt [1970], who pointed out that the same result could be achieved with only three subcells for each Fourier coefficient, as shown in fig. 10.5(*b*). Binary holograms based on this technique can also be produced.

10.4.   Phase randomization

The Fourier transforms of the wavefronts corresponding to most simple objects have very large dynamic ranges, because the coefficients of the dc and low-frequency terms have much larger moduli than those of the high-frequency terms. This results in nonlinearity because of the limited dynamic range of the recording medium (see section 6.5).

To minimize this problem, it is convenient, where the phase of the final reconstructed image is not important, to multiply the complex amplitudes at the original sampled object points by a random phase factor before calculating the Fourier transform. This is optically analogous to placing a diffuser in front of the object transparency and has the effect of making the magnitudes of the Fourier coefficients much more uniform, as shown in fig.

Fig. 10.5. Typical sub-cell arrangements and phasor diagrams for detour-phase holograms developed by (*a*) Lee [1970]; and (*b*) Burckhardt [1970].

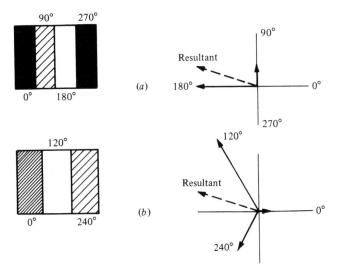

10.6. However, the reconstructed image is then modulated by a speckle pattern.

10.5. The kinoform

In the case where the object is diffusely illuminated, the magnitudes of the Fourier coefficients are relatively unimportant, and the object can be reconstructed using only the values of their phases. This led to the concept of a completely different type of hologram called a kinoform [Lesem, Hirsch & Jordan, 1969].

Fig. 10.6. Simulation of illumination through a diffuser (object with random phase): (a) the hologram; (b) the image [Lohmann & Paris, 1967].

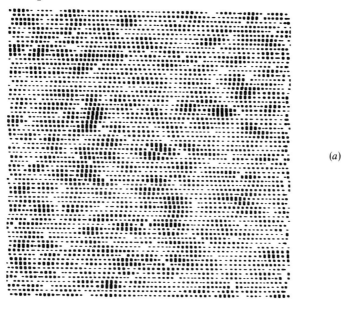

(a)

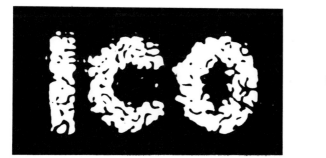

(b)

This is a computer generated hologram in which all the cells are completely transparent so that the moduli of all the Fourier coefficients are arbitrarily set equal to unity, and only the phase of the transmitted light is controlled in accordance with the phase of the computed Fourier coefficients. Thus, the amplitude transmittance t_{nm} of the cell corresponding to a Fourier coefficient with modulus $|o_{nm}|$ and phase ϕ_{nm} would be

$$t_{nm} = \exp{(i\phi_{nm})}. \tag{10.13}$$

However, to simplify recording, integral multiples of 2π radians are subtracted from the computed phases, so that they vary only between 0 and 2π over the entire kinoform.

To record the kinoform, the computed values of the phase ϕ_{nm} are encoded on a multilevel gray scale and used to control a photographic plotter which exposes a piece of film. The resulting master is then photographed once again, to reduce it to the final size, and bleached with a tanning bleach, to convert the gray levels to corresponding changes in optical thickness. With proper control of exposure and processing, the amplitude transmittance of the final kinoform can be made to conform closely to (10.13).

Kinoforms have the advantage that they can diffract all the incident light into the final image. However, to achieve this, care is necessary to ensure that the phase matching condition expressed by (10.13) is satisfied accurately [Kermisch, 1970]. Any error in the recorded phase shift results in light diffracted into the zero order (see the spot in the centre of figs. 10.7(a) and 10.7(c)) which can spoil the image.

10.6.   The referenceless on-axis complex hologram (ROACH)

This type of hologram makes use of multilayer colour film as a recording medium to obtain most of the advantages of the kinoform without its major disadvantages [Chu, Fienup & Goodman, 1973]. Different layers of the film are exposed selectively by light of different colours. When illuminated with light of a given colour, one layer of the film absorbs, while the other layers, which are effectively transparent, can cause phase shifts due to variations in the film thickness and refractive index. Thus, both the amplitude and phase of the transmitted beam can be controlled by a single element.

To record a ROACH which is to be illuminated finally with red light, the computed values of the moduli $|o_{nm}|$ of the Fourier coefficients are used, in the first instance, to control the brightness of a black-and-white CRT display. This is then photographed on to a reversal colour film, such as

Fig. 10.7. Photographs of two-dimensional images produced by kinoforms: (*a*) undermodulation (too short an exposure time); (*b*) correct exposure achieving good phase matching; (*c*) overmodulation (too long an exposure time) [Lesem, Hirsch & Jordan, 1969; copyright 1969 by International Business Machines Corporation, reprinted with permission].

(*a*)

(*b*)

(*c*)

Kodachrome II, using a red filter. In the next step, the computed values of the phases ϕ_{nm} of the complex Fourier coefficients are displayed in the same manner and photographed on the same frame of colour film using a blue–green transmitting filter.

After processing, the red absorbing layer of the film controls the amplitude of the transmitted red light so that it is proportional to the moduli $|o_{mn}|$ of the Fourier coefficients. The blue- and green-absorbing emulsions are transparent to red light, but they introduce phase shifts corresponding to the phases ϕ_{nm} of the Fourier coefficients, due to the thickness variations introduced by the blue–green exposure. In practice, the red-absorbing layer also introduces a phase shift proportional to the attenuation due to it. It is possible to compensate for this by subtracting from the blue–green exposure a component proportional to $|o_{nm}|$.

Since all the light is diffracted into a single image, the diffraction efficiency of the ROACH is very high. In addition, because both the amplitude and the phase of the object wave are encoded, the image quality is superior to that of the kinoform. The ROACH is also superior to the binary detour-phase hologram, because only one display spot is required for each Fourier coefficient, and quantization noise is negligible.

10.7.   Three-dimensional objects

The concept of computer holography can be generalized to a three-dimensional object [Waters, 1968; Lesem, Hirsch & Jordan, 1969; Brown & Lohmann, 1969]. Such an object is approximated by the sum of a number of equally spaced cross-sections perpendicular to the z axis. Problems can arise due to lines which are normally hidden by surfaces in front appearing in the image. To avoid this, it is necessary to sum the contributions to the object wave arising only from those points on the object that can be seen from a particular point on the hologram. Figure 10.8 shows a set of views of such an image from different angles, showing the resulting changes in parallax.

A completely different approach has been followed by King, Noll & Berry [1970]. Their technique is related to the techniques used in making holographic stereograms described in section 8.4 and has the advantage that it requires much less computer time. A computer is used to produce a series of perspective projections of the object, as seen from a number of angles in the horizontal plane. These are then optically encoded as a series of vertical strip holograms on a single plate. The real image formed by this composite hologram, when it is illuminated by the conjugate reference beam, is then used to produce an image hologram. Since this real image is

Fig. 10.8. Photographs of a three-dimensional image produced by a kinoform; the three geometrical figures are located in different spatial planes: (*a*) on-axis view; (*b*) and (*c*) off-axis views, showing changes in parallax [Lesem, Hirsch & Jordan, 1969; copyright 1969 by International Business Machines Corporation, reprinted with permission].

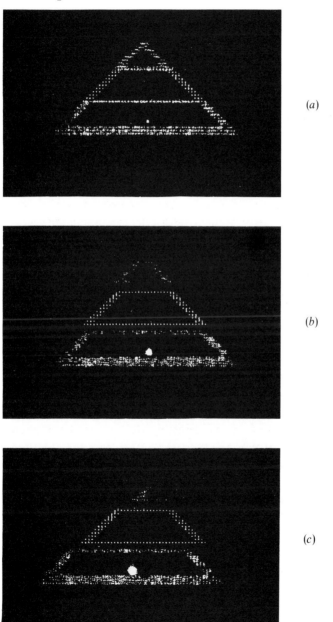

(*a*)

(*b*)

(*c*)

actually two-dimensional, it is located entirely in the plane of the final hologram, which can therefore be illuminated with white light to reconstruct a bright, almost achromatic image.

10.8.   Computer-generated interferograms

Problems can arise with detour-phase holograms when encoding wavefronts with large phase variations since, when the phase of the wavefront moves through a multiple of 2π rad, the two apertures near the crossover may overlap. This has led to an alternative approach to the production of binary holograms based on the fact that an image hologram of a wavefront that has no amplitude variations is essentially similar to an interferogram, so that the exact locations of the transparent elements in the binary hologram can be determined by solving a grating equation [Lee, 1974].

Different methods can then be used to incorporate information on the amplitude variations in the object wavefront into the binary fringe pattern [Lee, 1979]. In one, the two-dimensional nature of the Fourier transform hologram is used to record the phase information along the x direction, while the fringe heights in the y direction are adjusted to correspond to the amplitude. In another, the phase and the amplitude are recorded by the position and the width of the fringes along the direction of the carrier frequency, while in the third, the phase and amplitude of the object wave are encoded by the superposition of two phase-only holograms.

10.9.   Computer-generated holograms in optical testing

One of the main applications of computer-generated holograms is in interferometric tests of aspheric optical surfaces. Normally, such tests would require either an aspheric reference surface or an additional element, referred to as a null lens, which converts the wavefront produced by the element under test into a spherical or plane wavefront. However, Pastor [1969] and Snow & Vandewarker [1970] showed that a hologram of the reference surface or the null lens could be used instead, and it was not long before MacGovern & Wyant [1971] showed that a computer-generated hologram could be used instead of one recorded in the conventional manner.

An experimental setup using a Twyman–Green interferometer in conjunction with a computer-generated hologram to test an aspherical mirror is shown in fig. 10.9 [Wyant & Bennett, 1972]. The hologram used is a binary representation of the interferogram that would be obtained if the wavefront from an ideal aspheric surface were to interfere with a tilted plane wavefront, and is placed in the plane in which the mirror under test is imaged. The superposition of the actual interference fringes and the computer-generated hologram produces a moiré pattern which gives the deviation of the actual wavefront from the ideal computed wavefront.

The contrast of the moiré pattern is improved by spatial filtering. For this, the hologram is reimaged with an appropriately placed small aperture in the focal plane of the reimaging lens. This aperture passes only the transmitted wavefront from the mirror under test and the diffracted wavefront produced by illuminating the hologram with a plane wavefront. This is possible if, in producing the computer-generated hologram, the slope of the plane reference wavefront is greater than the maximum slope of the aspheric wavefront along the same direction.

Typical fringe patterns obtained with an aspherical surface, with and without the computer-generated hologram, are shown in fig. 10.10.

If good results are to be achieved, a number of potential sources of error must be kept in mind. The first of these is the effect of quantization. If there are N resolvable points across the diameter of the hologram, any plotted point may be displaced from its proper position by a distance equal to $1/2N$ of the diameter. However, for the diffracted spectra not to overlap, the hologram must have a carrier frequency of $3s$ where s is the highest spatial

Fig. 10.9. Twyman–Green interferometer modified to use a computer-generated hologram to test an aspherical mirror [Wyant & Bennett, 1972].

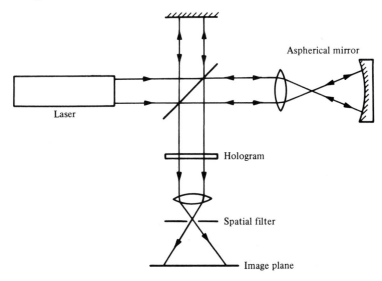

frequency (fringes per diameter) in the uncorrected interference pattern. This means that the fringe frequency in the hologram can vary from a minimum of $2s$ to a maximum of $4s$. If there are $4s$ fringes across the hologram, an error in the fringe position of $1/2N$ would correspond to a wavefront error (expressed as a fraction of a fringe)

$$\Delta W = 2s/N. \tag{10.14}$$

Accordingly, for a maximum wavefront error of $\lambda/4$ the slope of the wavefront under test should not exceed

$$s = N/8, \tag{10.15}$$

fringes per diameter.

Another source of error is plotter distortion. If e is the maximum distortion introduced by the plotter over a plotting surface of diameter D, the corresponding wavefront error has a maximum value

$$\Delta W = (e/D)4s. \tag{10.16}$$

For this not to exceed $\lambda/4$,

$$e < D/16s. \tag{10.17}$$

A method to minimize such errors [Fercher, 1976] is to use a hologram which contains two superposed structures, one corresponding to the object wave, while the other is a linear grating. Plotter errors, being common to both, are eliminated.

Errors in the size of the hologram introduce a radial shear between the reference wave and the wavefront under test. The wavefront error due to

Fig. 10.10. Interferometer tests of an aspherical wavefront having a maximum slope of thirtyfive waves per radius and a maximum departure of nineteen waves from a reference sphere, (a) with, and (b) without a computer-generated hologram [Wyant & Bennett, 1972].

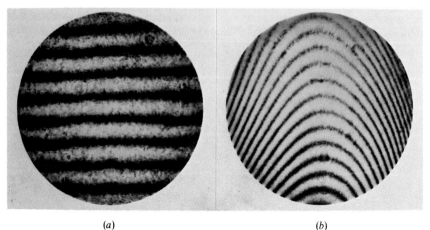

(a)   (b)

this is

$$\Delta W = s(\Delta d/d), \tag{10.18}$$

where d is the nominal diameter of the hologram.

If this wavefront error is not to exceed $\lambda/4$,

$$\Delta d < d/4s. \tag{10.19}$$

Improper positioning of the hologram also introduces a shear. Thus, a sideways translation Δx results in a lateral shear which can cause a wavefront error

$$\Delta W = s(\Delta x/d), \tag{10.20}$$

where s is now the slope of the test wavefront along the direction of shear. The condition for this wavefront error to be less than $\lambda/4$ is, therefore,

$$\Delta x < d/4s. \tag{10.21}$$

Different methods for plotting the fringes in such holograms have been discussed by Birch & Green [1972]. However, none of these are satisfactory when testing surfaces with large deviations from a sphere because of the relatively high fringe frequencies required. While Sirohi, Blume & Rosenbruch [1976] have shown how the computation time involved can be reduced by merging two methods, a more common procedure is to use a combination of a null lens and a computer-generated hologram [Faulde, Fercher, Torge & Wilson, 1973; Wyant & O'Neill, 1974]. The design of the null lens can be quite simple, since it need only reduce the residual aberrations to a level which can be handled by the computer-generated hologram. Other approaches to minimize the number of fringes in the hologram have involved the use of on-axis holograms [Ichioka & Lohmann, 1972; Mercier & Lowenthal, 1980] and aberration-balancing techniques [Yatagai & Saito, 1978]. An entirely different technique is the dual computer-generated hologram described by Yatagai & Saito [1979]. This consists of apertures whose positions are modulated along two different directions, allowing the simultaneous generation of two diffracted wavefronts whose sum and difference can be obtained by suitably positioned spatial filters.

Computer-generated holograms are now used as a routine to test aspheric surfaces: see for example the reviews by Lukin & Mustafin [1979] and Loomis [1980]. An active field of research has been the development of improved plotting routines and the application of techniques such as electron-beam recording, using layers of photoresist coated on optically worked substrates, for the production of computer-generated holograms of very high quality [Biedermann & Holmgren, 1977; Leung, Lindquist & Shepherd, 1980].

11

Special techniques

11.1. Polarization recording

With normal holographic techniques, the amplitude and phase of the object wavefront are recorded accurately, but information on its state of polarization is lost. The polarization of the reconstructed wave is determined by the polarization of the light used to illuminate the hologram.

11.1.1. Orthogonally polarized reference beams

Two basic methods for recording the state of polarization of the object wave have been described. The first method was proposed by Lohmann [1965a] and subsequently demonstrated by Bryngdahl [1967]. The experimental arrangement for this method, which is shown in fig. 11.1, uses two orthogonally polarized reference waves which interfere with the corresponding polarized components of the light from the object, so that two holograms are recorded on the same plate. After processing, when the plate

Fig. 11.1. Experimental arrangement for recording the state of polarization of the object wave using Lohmann's method [Gåsvik, 1975].

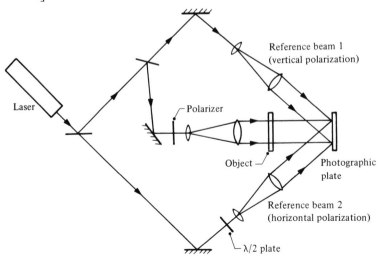

is illuminated once again with the same reference beams, it yields two superimposed images which reproduce the polarization of the object wave. Care must be taken, of course, to arrange the setup so that the cross-talk images formed by diffraction of each of the reference beams by the hologram formed with the orthogonally polarized component do not overlap the desired image.

11.1.2. Coded reference beams

The other method, which is due to Kurtz [1969], uses a single reference beam in which, as shown in fig. 11.2, an opal glass diffuser is inserted. The light transmitted by this diffuser is depolarized and the complex amplitudes of the two orthogonally polarized components of the reference beam at any point exhibit little or no correlation. Because of this, the orthogonally polarized components of the object wave are effectively encoded by different random wavefronts [see section 13.3].

11.1.3. Conditions for accurate recording

These methods have been analysed in detail by Windischbauer *et al.* [1973] and by Gåsvik [1975]. The latter has shown that several conditions must be satisfied to ensure accurate reproduction of the state of polarization of the object wave.

A problem common to both methods is that the component of the electric vector parallel to the plane of incidence produces a hologram with lower modulation, and hence lower diffraction efficiency, than the component normal to the plane of incidence (see section 5.3). This difference can be brought within acceptable limits if the angle between the object beam and the reference beam(s) is made fairly small.

In addition, with Lohmann's method, inaccurate repositioning of the hologram after processing, as well as variations in the optical paths of the

Fig. 11.2. Setup for recording the state of polarization of the object wave using Kurtz's method [Gåsvik, 1975].

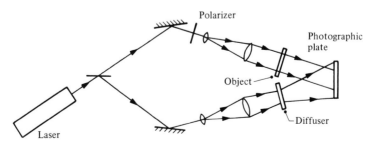

two reference beams due to temperature gradients, will result in changes in the polarization state of the reconstructed wave. These errors can be minimized by using a modified optical system in which the two reference beams run quite close to each other.

The latter problem is not present in Kurtz's method, but, in this case, the processed hologram must occupy exactly the same position in which it was exposed if it is to reconstruct an image (see section 13.3). Normally, this implies that the hologram must be processed *in situ*. In addition, the reconstructed image is quite noisy due to the presence of a uniform background of scattered light. With a transparent object, this can be minimized by inserting an aperture in the back focal plane of the imaging lens. However, there is always a considerable loss of light due to the diffuser.

Both methods require the hologram to be illuminated with the same wavelength as that used to record it. With Lohmann's method a change in the wavelength results in a spatial variation in the state of polarization, while with Kurtz's method there is no reconstruction at all.

Techniques for polarization recording have mainly found application, so far, in holographic photoelasticity (see section 15.2).

11.2.   Holography with incoherent illumination

It is an interesting fact that image-forming holograms can be recorded even with incoherent illumination. While all the techniques developed so far suffer from some limitations, they do point the way towards several interesting possibilities.

Mertz & Young [1962] were the first to show that the only necessary condition for recording a hologram was that each point on the object should produce a two-dimensional pattern which uniquely encoded its position and its intensity. For this, they used a mask in the form of a Fresnel zone plate, so that each object point cast a shadow of this form on the film. When illuminated, each Fresnel zone plate recorded on the film produced an image of the corresponding object point.

While this geometrical approach is applicable to very short wavelengths, such as x-rays, it breaks down at longer wavelengths. Accordingly, early methods of holography with spatially incoherent illumination used interference between two wavefronts derived from the object. This is possible with a shearing interferometer. With incoherent illumination, interference takes place only between the waves derived from the same point on the object, which are mutually coherent. As many independent, superimposed interference patterns are recorded as there are independent points on the object. Paradoxically, this lack of spatial coherence between the various points on the object is an essential condition for this method to work. Of course, to obtain interference fringes of good contrast it is necessary for the coherence length of the light to be much greater than the path differences involved.

To encode each object point uniquely, the interferometer must produce a shear which is a function of the position of the object point. This is possible with either a rotational or a radial shear; in both cases the shear is directly proportional to the position vector of the point in the object plane.

11.2.1. Rotational shear systems

For plane objects it is possible to use an interferometer which introduces a rotational shear of $180°$ between the two waves [Stroke & Restrick, 1965; Worthington, 1966]. If we consider an object point O with coordinates (x_0, y_0) the optical system of the interferometer produces two images of this point located at (x_0, y_0) and $(-x_0, -y_0)$ in the object plane. The interference pattern produced by these two virtual sources is a series of equally spaced, straight fringes at right angles to the line joining the two images. The intensity distribution in the hologram plane can then be written, apart from a constant factor, as

$$\Delta I(\xi, \eta) = I(x_0, y_0)\{1 + \cos[2\pi(x_0\xi + y_0\eta)]\}, \tag{11.1}$$

where $I(x_0, y_0)$ is the intensity of the object point, z_0 is the distance from the object to the hologram and ξ and η are coordinates in the hologram plane defined by the relations $\xi = 2x/\lambda z_0$ and $\eta = 2y/\lambda z_0$. The total intensity in the hologram plane is then merely the sum of these contributions, so that

$$I(\xi, \eta) = \int\!\!\!\int_{-\infty}^{\infty} I(x_0, y_0)\{1 + \cos[2\pi(x_0\xi + y_0\eta)]\} \mathrm{d}x_0 \mathrm{d}y_0. \tag{11.2}$$

This is, apart from a constant background, the two-dimensional Fourier transform of the intensity distribution in the object. Accordingly, if the hologram is illuminated with a spatially coherent monochromatic wave, it will reconstruct, in the far field, two images of the object positioned symmetrically about the optical axis. Variation of the rotational shear permits varying the scale of the interference pattern and, hence, of the reconstructed image [Lowenthal, Serres & Froehly, 1969].

11.2.2. Radial shear systems

Rotational shear systems cannot give a three-dimensional image, because the interference pattern produced by an object point is always a series of straight lines irrespective of its distance from the hologram plate. To reproduce depth information, the shear introduced between the wavefronts must have a component parallel to the axis so that the fringe pattern is equivalent to a Fresnel transform of the scene [Lohmann, 1965b]. Cochran [1966] used the triangular-path interferometer shown in fig. 11.3 containing an afocal system which produces two wavefronts with a radial shear [Hariharan & Sen, 1961].

If, in this interferometer, the planes A and B are located at distances from L_1 and L_2 equal to their respective focal lengths, the paths of the two beams, which traverse the same circuit in opposite directions, can be developed as shown in fig. 11.4. Two images of A are therefore formed at B with magnifications of $(1/\alpha) = -f_2/f_1$ and $(1/\beta) = -f_1/f_2$, respectively. An object point $O(x_0, y_0)$ gives rise to two images $O'(\alpha x_0, \alpha y_0)$ and $O''(\beta x_0, \beta y_0)$.

It can be shown (see Appendix A2.1) that the optical path from an object point (x_0, y_0, z_0) to a point (x,y) in the reference plane is given by the relation

$$r(x, y) = r_0 - (x_0 x + y_0 y)/r_0 + (x^2 + y^2)/2r_0 \\ - (x_0 x + y_0 y)^2/2r_0^3 - \ldots, \tag{11.3}$$

where $r_0^2 = x_0^2 + y_0^2 + z_0^2$. Accordingly, the difference in the optical paths from the images O' and O'' to a point $P(x, y)$ in the hologram plane, which we assume to be at a distance z_0, is

$$r'(x, y) - r''(x, y) = -(\alpha - \beta)(x_0 x + y_0 y)/r_0 \\ + (\alpha^2 - \beta^2)(x^2 + y^2)/2r_0 \\ - (\alpha^2 - \beta^2)(x_0 x + y_0 y)^2/2r_0^3 - \ldots. \tag{11.4}$$

The common factor $(\alpha - \beta)$ is merely a constant of proportionality which

Fig. 11.3. Optical system used for holography with incoherent illumination, in which interference takes place between two waves with a radial shear [Cochran, 1966].

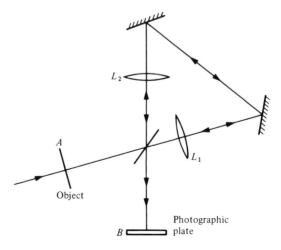

determines the scale of the interference pattern on the hologram, and we can equate it to unity. We can then simplify (11.4) further if we set

$$x_1 = x_0/(\alpha + \beta),$$ (11.5)

$$y_1 = y_0/(\alpha + \beta),$$ (11.6)

$$z_1 = z_0/(\alpha + \beta),$$ (11.7)

$$r_{10} = r_0/(\alpha + \beta).$$ (11.8)

With these substitutions, the difference in the optical paths is

$$\begin{aligned} r'(x, y) - r''(x, y) = &-(x_1 x + y_1 y)/r_{10} \\ &+ (x^2 + y^2)/2r_{10} \\ &- (x_1 x + y_1 y)^2/2r_{10}^3 - \dots . \end{aligned}$$ (11.9)

A comparison of (11.9) with (11.3) shows that (11.9) can be written as

$$r'(x, y) - r''(x, y) = r_1(x, y) - r_{10},$$ (11.10)

where $r_1(x, y)$ is the distance from a point (x_1, y_1, z_1) to a point (x, y) in the reference plane, and, from (11.5)–(11.8), $r_{10}^2 = x_1^2 + y_1^2 + z_1^2$. For any given object point, r_{10} is a constant, and the path difference between the wavefronts defined by the right hand side of (11.10) corresponds to that between a sphere of radius r_{10} with its centre at (x_1, y_1, z_1) and the reference plane.

If then the intensities of the two images are equalized by a proper choice of the reflectance and transmittance of the beamsplitter, the intensity distribution in the resulting interference pattern is

$$I(x, y) = I_0\{1 + \cos(2\pi/\lambda)[r_1(x, y) - r_{10}]\}.$$ (11.11)

Fig. 11.4. Equivalent optical systems for the two beams circulating in opposite directions in a triangular path interferometer.

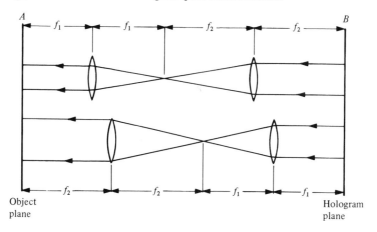

The hologram is, therefore, a Fresnel zone pattern. When illuminated with collimated light, it reconstructs an image of the object point at (x_1, y_1, z_1), as well as a conjugate image at $(-x_1, -y_1, -z_1)$.

With this setup, it is possible to make a hologram that reconstructs a three-dimensional image. In addition, the magnification of the image can be varied by varying the radial shear. However, undistorted depth information can be obtained only for a magnification of -1 [Bryngdahl & Lohmann, 1970a].

11.2.3. Diffraction efficiency

The main drawback of holograms recorded with spatially incoherent illumination is that their diffraction efficiency drops off rapidly as the complexity of the object increases. As is apparent from (11.11), the intensity in the hologram plane due to a single object point is the sum of a uniform bias and a spatially varying term. With a number of incoherently illuminated points, the intensity is merely the sum of the contributions due to the individual points. It can be shown [Cochran, 1966] that, for an object consisting of N independently radiating points, the contrast of the fringes in the hologram is proportional to $(1/N)$, while the diffraction efficiency is proportional to $(1/N^2)$. Various methods have been proposed to minimize this loss. One proposal involves eliminating the bias term by time-modulation of one of the beams [Kozma & Massey, 1969], while another is to form an image in the vertical direction and a series of one-dimensional holograms in the horizontal direction [Bryngdahl & Lohmann, 1968b], each of which receives contributions from only a limited number of points.

11.2.4. Achromatic systems

The temporal coherence required to record a hologram can be reduced by using an interferometer that yields achromatic fringes [Leith & Upatnieks, 1967]. As shown in fig. 11.5, a grating located in the plane P_1 illuminated by a collimated beam is imaged at unit magnification into the plane P_4. The diffracted waves are focused at P_2, the back focal plane of the lens, where a stop passes only the zero order and one of the first-order diffracted beams, which then form an image of the grating at P_4. A scattering object, such as a transparency, is located at P_3 in the zero-order beam, while the diffracted beam serves as the reference beam.

It is obvious that the interference fringes formed at P_4 may also be regarded in this case as images of the grating rulings. Accordingly, the temporal coherence requirements are determined mainly by the characteristics of the object. Good holograms of transparencies can be recorded with a high-pressure mercury-vapour lamp illuminating a 500 μm pinhole.

For the system to be truly achromatic it is necessary for the diffraction pattern of the object in the plane of the hologram to be almost the same for all wavelengths. This is possible if the object is also imaged at, or near, the hologram plane. Bryngdahl & Lohmann [1970*b*] showed that it was then possible to record a hologram with a xenon arc lamp as the source.

A more efficient system is possible with an interferometric arrangement that uses diffraction gratings both to split and recombine the beams [Leith & Chang, 1973]. Such an interferometer can be made achromatic and can, in addition, use an extended source. The sizes of the object and the hologram can then be made relatively large without altering the coherence requirements.

In a holographic setup of this type shown in fig. 11.6 [Chang, 1973], two identical gratings G_1 and G_2 are used to split and recombine the beams. The object transparency O is inserted in one of the beams at a distance d_0 from the hologram plane H. An interesting feature of this arrangement is that its modulation transfer function can be varied by changing the position of one of the gratings. White light from a relatively broad source can be used both to record holograms and to view the reconstructed image.

A more general analysis of a three-grating interferometer used as a hologram-forming system has been presented by Chang & Leith [1979]. Such a system can reduce the coherence requirements for off-axis holography to less than those for in-line holography, and can be adjusted to produce low-pass, bandpass or high-pass characteristics.

Fig. 11.5. Grating achromatic-fringe interferometer used to record holograms with a source of appreciable spectral bandwidth [Leith & Upatnieks, 1967].

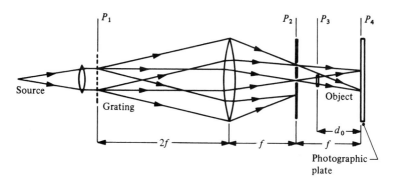

11.3.   Hologram copying

There are quite a few situations in which many identical holograms are required. A considerable amount of time and labour can then be saved if copies are made from a single hologram of the original object. Several techniques are available for this purpose (see, for example, the review by Vanin [1978]).

Where the number of copies required is large, mechanical copying methods which make use of the surface relief of the original hologram are suitable. Copies are made on a thermoplastic by conventional pressing techniques, using either an electroformed master [Bartolini, Hannan, Karlsons & Lurie, 1970] or the hologram itself as the master [Vagin & Shtan'ko, 1974]. Alternatively, a cold-setting polyester resin can be used to make replicas by casting [Trukhmanova & Denisyuk, 1977].

High-quality relief masters can be produced with a photoresist (see section 7.3). However, good relief masters can also be obtained at lower spatial frequencies with silver-halide photographic materials. In such cases, the use of a tanning developer and a tanning bleach [Altman, 1966; Smith, 1968; Butusov & Ioffe, 1976] gives increased relief.

11.3.1.   Optical methods

Holograms can also be copied by optical methods. These have the advantage that they do not require special equipment and permit the use of a wide range of materials which can be handled easily in the laboratory. One method is to illuminate the hologram with the conjugate of the reference beam used to record it. The wave reconstructed by the hologram can then be used with another reference wave to record a second-generation hologram. This technique has the advantage of great flexibility. It is

Fig. 11.6. Grating interferometer used to record holograms with a white-light source [Chang, 1973].

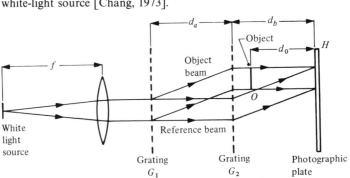

possible to produce a copy that reconstructs an orthoscopic real image [Rotz & Friesem, 1966] and to control the spatial frequency spectrum, the viewing angle and the brightness of the reconstructed image. This copying technique can even be used to produce copies with improved diffraction efficiency from a hologram with relatively low diffraction efficiency [Palais & Wise, 1971]. It is also possible to produce a number of reflection holograms from a transmission hologram of the object [Vanin, 1978; Růžek & Fiala, 1979].

The requirements on coherence and stability for this technique are very nearly as exacting as for recording the original hologram. Accordingly, a much simpler method which is widely used is to 'contact print' the original on to another photosensitive layer [Harris, Sherman & Billings, 1966]. For diffraction effects to be negligible, the separation Δz between the hologram and the copy must satisfy the condition

$$\Delta z < \Lambda^2/4\lambda, \qquad (11.12)$$

where Λ is the average spacing of the hologram fringes. Accordingly, true contact printing is possible only with thin holograms having a low spatial-carrier frequency.

11.3.2. Coherence requirements for 'contact printing'

In most cases, the spatial-carrier frequency of the hologram is high enough that the separation between the hologram and the copy in 'contact printing' cannot be neglected. It is then necessary to take into account diffraction effects at the hologram [Nassenstein, 1968a]. Consider a thin hologram and assume that, as shown in fig. 11.7, the hologram H_1 and the

Fig. 11.7. Coherence requirements for copying a hologram by 'contact printing'.

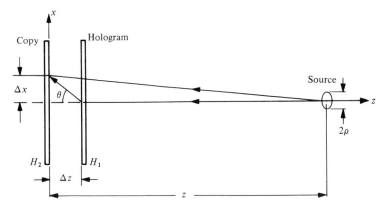

photographic plate H_2 used to copy it are separated by a small distance Δz and illuminated by a uniform extended source of diameter 2ρ located at a comparatively large distance z. In this case, what is recorded on the photographic plate is actually the interference pattern formed by the light diffracted by the hologram and the light transmitted by it. Hence, for a satisfactory copy to be obtained, the coherence of the illumination must be adequate to produce interference fringes of high visibility at H_2.

If the hologram diffracts light at an angle θ, interference takes place between rays originally separated by a distance Δx where

$$\Delta x = \Delta z \tan \theta. \tag{11.13}$$

With an extended source, the degree of spatial coherence of the interfering beams is then given, from the van Cittert–Zernike theorem, by the normalized Fourier transform of the intensity distribution over the source (see Appendix A3.3), so that

$$|\gamma(v)| = 2J_1(v)/v, \tag{11.14}$$

where $v = 2\pi\rho\Delta x/\lambda z$. If we assume that $\Delta z < 100\ \mu m$, and $\tan\theta < 0.3$, the degree of spatial coherence with a source of diameter $2\rho = 2$ mm, at a distance of 1 m would be approximately 0.98, which is quite satisfactory.

It is also apparent from fig. 11.7 that the path difference between the beams is $\Delta x \sin\theta$. This must be less than Δl, the coherence length, so that

$$\Delta l > \Delta x \sin\theta, \tag{11.15}$$

or, from (11.13),

$$\Delta l > \Delta z \tan\theta \sin\theta. \tag{11.16}$$

If, as before, $\Delta z < 100\ \mu m$ and $\tan\theta < 0.3$, $\Delta l > 10\ \mu m$.

It is apparent from (11.13) and (11.16) that the coherence requirements for hologram copying by 'contact printing' are much less stringent than for recording a hologram, and that, in many cases, satisfactory copies of holograms can be made by this technique even with a high-pressure mercury vapour lamp.

In addition, to obtain a copy with good diffraction efficiency, the amplitudes of the transmitted wave and the diffracted wave from H_1 must be comparable. This condition can usually be met with an amplitude transmission hologram by adjusting the exposure so that the amplitude transmittance of the hologram is less than 0.1. An index-matching fluid can be used between the two plates to minimize spurious interference fringes formed by reflection between the emulsion surfaces.

11.3.3. Image doubling

Problems of spatial and temporal coherence can be eliminated completely by the use of a laser source, so that the separation of H_1 and H_2 is no longer critical. However, large separations result in a copy which generates double images [Brumm, 1967]. The separation of the twin primary and conjugate images is, in general, a function of the curvature of the reference and signal waves as well as the separation of H_1 and H_2 [Vanin, 1978]. For a plane reference wave, the separation of the twin images is twice the distance between H_1 and H_2, while for a Fourier hologram, all four images are at the same distance.

11.3.4. Volume holograms

The problem of image doubling is much less serious when copying volume transmission holograms, since the angular selectivity of the hologram can be made high enough to ensure that the amplitude of the conjugate reconstructed wave is negligible [Belvaux, 1967; Sherman, 1967]. However, with normal illumination, effects due to the thickness of the emulsion can be quite pronounced, resulting in low modulation in certain planes and copies of poor quality [Landry, 1967; Kaspar, 1974]. The MTF of the copying process has also been studied by Suhara, Nishihara & Koyama [1975], who have shown that for best results the copying setup must duplicate the direction, curvature and wavelength of the original reference wavefront.

Volume reflection holograms can also be copied by interchanging the positions of H_1 and H_2 in fig. 11.7 [Belvaux, 1967; Kurtz, 1968]. In this case the illuminating beam passing through the unexposed emulsion acts as the reference beam, while the light diffracted back from H_1 constitutes the object beam for the second hologram. If the amplitudes of the two beams are to be comparable, the diffraction efficiency of H_1 must be quite high. For this it is necessary to duplicate the direction, curvature and wavelength of the reference beam used in recording H_1. In addition, it is also necessary to see that H_1 is processed to avoid shrinkage of the photographic emulsion [Zemtsova & Lyakhovskaya, 1976]. With these precautions, copies of quite good quality can be made. This method works best with holograms of bright, polished objects.

12
Applications in imaging

12.1. Holographic microscopy

As mentioned in Chapter 1, holographic imaging was originally developed in an attempt to obtain higher resolution in microscopy. Equations (3.20) and (3.21) show that it is possible to obtain a magnified image if different wavelengths are used to record a hologram and reconstruct the image, or if the hologram is illuminated with a wave having a different curvature from the reference wave used to record it. However, neither of these techniques has found much use, in the first instance because of the limited range of laser wavelengths available, and, in the second, because of problems with image aberrations [Leith & Upatnieks, 1965; Leith *et al.*, 1965].

The most successful applications of holography to microscopy have been with systems in which holography is combined with conventional microscopy. In one approach, a hologram is recorded of the magnified real image of the specimen formed by the objective of a microscope, and the reconstructed image is viewed through the eyepiece [van Ligten & Osterberg, 1966]. While this offers no advantages for ordinary subjects, it is extremely useful for phase and interference microscopy [Snow & Vandewarker, 1968]. In another, a hologram is recorded of the object, and the reconstructed real image is examined with a conventional microscope. This technique is particularly well adapted to the study of dynamic three-dimensional particle fields, as described in the next section.

12.2. Particle-size analysis

Measurements on moving microscopic particles distributed throughout an appreciable volume are not possible with a conventional optical system. This is because a microscope which can resolve particles of diameter d has a limited depth of field

$$\Delta z \approx d^2/2\lambda. \tag{12.1}$$

Holography permits storing a high-resolution, three-dimensional image of the whole field at any instant. The stationary image reconstructed by the

174

hologram can then be examined in detail, throughout its volume, with a conventional microscope (see, for example, the reviews by Thompson [1974] and Trolinger [1975a]).

In-line holography can be used for such studies wherever a sufficient amount of light (> 80 per cent) is directly transmitted to serve as a reference beam. This permits a very simple optical system, which is also economical of light. However, a distinction must be made between such an in-line hologram of a particle field and a Gabor hologram. Because of the small diameter d of the particles, the distance z of the recording plane from the particles easily satisfies the far-field condition $z \gg d^2/\lambda$ (see Appendix A2.2), so that the diffracted field due to the particle is its Fraunhofer diffraction pattern. The hologram formed by the interference of the diffracted light and the directly transmitted light is, therefore, a Fraunhofer hologram (see section 2.6).

The permissible exposure time for recording a hologram of a moving particle field depends on the velocity of the particles. For size analysis, a useful criterion is that the particle should not move by more than a tenth of its diameter during the exposure. Typically, a 10 μm particle moving with a velocity of 1 m s^{-1} would require an exposure time less than 10^{-6} s. Suitable sources are either a pulsed ruby laser or, where a higher repetition rate is necessary, a frequency-doubled Nd:YAG laser.

Again, to give a satisfactory reconstructed image of a spherical particle, the hologram must record the central maximum and at least three side-lobes of its diffraction pattern. This would correspond to waves travelling at a maximum angle

$$\theta_{max} = 4\lambda/d, \tag{12.2}$$

to the directly transmitted wave, and, hence, to a maximum fringe frequency of $4/d$, which is independent of the values of λ and z. Accordingly, for a 10 μm diameter particle, the recording material should have an MTF which extends beyond 400 mm^{-1}.

Films with lower resolution, which are faster and cheaper, can be used if a hologram is recorded of a magnified image of the particles. For this, the particle field is imaged near the hologram plane, as shown in fig. 12.1, with a telescopic system having a magnification of about 5. This leaves the reference beam parallel and gives constant magnification over the whole depth of the field.

It is also apparent from this discussion that the depth of field is limited mainly by the dimensions of the recording material. From (12.2) it follows that x, the half-width of the hologram, must be greater than $4z\lambda/d$. Hence, the maximum depth of the field over which the required resolution can be maintained is given by the relation

$$\Delta z_{max} = xd/4\lambda. \qquad (12.3)$$

For a 10 μm diameter particle and a film halfwidth of 1 cm, the depth of field is approximately 500 times that given by (12.1) for a conventional microscope.

To reconstruct the image, the hologram is illuminated with a collimated beam from a He–Ne laser. With normal processing, negative images are formed, but this is no problem for many technical applications. Two images are reconstructed, at equal distances z from the hologram and on opposite sides of it. However, with a Fraunhofer hologram, the contribution of one image in the plane of the other is essentially a constant and, therefore, does not degrade the image significantly.

Off-axis holography has also found considerable use in particle size analysis [Wuerker & Heflinger, 1970]. Apart from the fact that the two reconstructed images are completely separated and positive images are obtained, its major advantage is that the reference beam need not traverse the sample volume. This makes it possible to study samples with poor transmission in reflected light. Its main disadvantages are the need for a separate reference-beam path and more stringent demands on the coherence of the illumination and the resolution of the recording medium.

The applications of holographic particle-size analysis include studies of fog droplets, dynamic aerosols and marine plankton. Another significant area of application has been in bubble-chamber photography, where a depth of field of over 2 m has been achieved. Double-exposure holography with a suitable time interval between the light pulses has also been found useful to measure the velocity spectrum. A rapid method of analysis is based upon observations in the Fraunhofer plane on reconstruction [Ewan, 1979]. The diffraction pattern of the original particle field is modulated by

Fig. 12.1. In-line Fraunhofer holographic system for particle-size analysis.

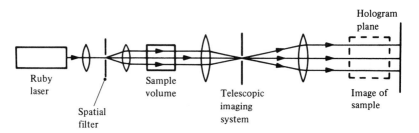

fringes whose spacing can be shown to be related to the velocity of the particles.

12.3. Imaging through moving scatterers

Holography can be used quite effectively to produce an image of a stationary object masked by moving scatterers [Stetson, 1967b; Spitz, 1967]. This is possible because light scattered from a moving particle has its frequency shifted by the Doppler effect. Accordingly, if a hologram is recorded of the object with a reference beam which does not pass through the scattering medium, only the directly transmitted light contributes to the formation of the hologram. The light scattered by the moving particles cannot interfere with the reference beam and merely adds a constant exposure to the hologram plate. This has little effect on the reconstructed image as long as the recording is linear [Hamasaki, 1968], though there is a decrease in diffraction efficiency when the amount of scattered light is large.

12.4. Imaging through distorting media

Another situation where holography permits forming an undistorted image of an object is where the reference wave can be made to undergo the same distortion as the object wave [Goodman, Huntley, Jackson & Lehmann, 1966]. This is possible if the distorting medium is very thin and lies close to the hologram plane, or if the angular separation of the two waves is very small.

Let the complex amplitudes of the object wave and the reference wave incident on the medium be $o(x, y)$ and $r(x, y)$ respectively. If the distorting medium modulates only the phase of an incident wave, its amplitude transmittance can be written as $\exp[-i\phi(x, y)]$. The complex amplitudes of the object wave and the reference wave at the hologram are then $o(x, y)\exp[-i\phi(x, y)]$ and $r(x, y)\exp[-i\phi(x, y)]$. Assuming linear recording as defined by (2.2), the amplitude transmittance of the hologram is

$$\begin{aligned} \mathbf{t}(x, y) &= \mathbf{t}_0 + \beta T |r(x, y)\exp[-i\phi(x, y)] \\ &\quad + o(x, y)\exp[-i\phi(x, y)]|^2, \\ &= \mathbf{t}_0 + \beta T |r(x, y) + o(x, y)|^2. \end{aligned} \qquad (12.4)$$

Since the phase variations $\phi(x, y)$ due to the distorting medium have been eliminated, an undistorted image is formed when the hologram is illuminated by the undistorted reference wave $r(x, y)$.

12.5. Correction of aberrated wavefronts

Holography can also be used, within certain limits, to recover an image of an object, unaffected by lens aberrations [Upatnieks, Vander Lugt

& Leith, 1966] or by the presence of an aberrating or diffusing medium in the optical path [Kogelnik, 1965; Leith & Upatnieks, 1966].

As shown in fig. 12.2(a), the object is illuminated with coherent light and a hologram is recorded of the aberrated image wave using a collimated reference beam. Let the complex amplitude of the aberrated object wave in the hologram plane be

$$o(x) = |o(x)| \exp [-i\phi(x)]. \qquad (12.5)$$

In the reconstruction step, the hologram is illuminated, as shown in fig. 12.2(b), by the conjugate to the original reference wave. The hologram then reconstructs the conjugate to the original object wave, whose complex amplitude in the hologram plane can be written, apart from a constant factor as

$$o^*(x) = |o(x)| \exp [i\phi(x)]. \qquad (12.6)$$

Fig. 12.2. Holographic system for imaging through an aberrating medium: (a) hologram recording; (b) image reconstruction.

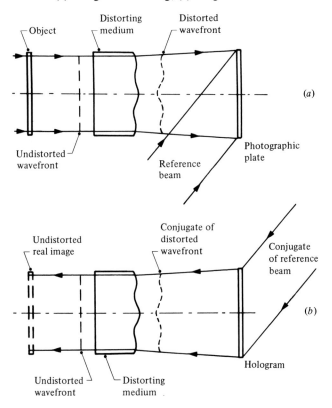

This wave has exactly the same phase errors as the original object wave, except that they are of the opposite sign. Hence, when this wave propagates back through the optical system, these phase errors cancel out exactly, so that the wave emerging from it is the undistorted object wave. As a result, a diffraction-limited real image of the object is formed in its original position.

Generation of an undistorted image in real time is possible with a hologram recording medium such as $Bi_{12}SiO_{20}$ (BSO) (see section 7.7.2). A setup for this is shown in fig. 12.3 [Huignard, Herriau, Aubourg & Spitz, 1979]. Continuous generation of the phase-conjugate wavefront is possible using an Ar^+ laser ($\lambda = 488$ nm) at power levels around 6 W/m^2. This can also be looked at as an example of four-wave mixing in the BSO crystal.

12.6.   High-resolution projection imaging

High-resolution imaging of photolithographic masks is an essential step in the production of semiconductor devices, but even the best photographic lenses have limitations on the resolution and the field they can cover. Holographic imaging has considerable potential for such work [Beesley, Foster & Hambleton, 1968]. However, there are problems in achieving the performance which should be possible in theory, due to degradation of the image by speckle, noise introduced by the recording material and the need for exact alignment of the hologram with the illuminating beam [Champagne & Massey, 1969].

These problems can be minimized if the object plane and the hologram

Fig. 12.3. Real-time imaging through an aberrating medium [Huignard *et al.*, 1979].

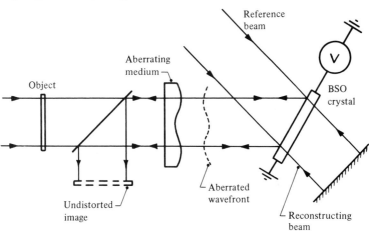

Reference beam

Aberrating medium

Object

BSO crystal

Aberrated wavefront

Undistorted image

Reconstructing beam

plane are very close to each other. An arrangement for this purpose described by Stetson [1967a] is shown in fig. 12.4. In this, a prism is cemented on to the back of the hologram plate. This permits a collimated reference beam to be brought in, so that it is incident at the outer face of the emulsion at an angle greater than the critical angle and is totally reflected. The object transparency, which is separated from the emulsion layer by a layer of tape, is illuminated from above, light transmitted through the emulsion being reflected out sideways by the prism.

In this setup, two holograms are recorded in the same emulsion. One is a reflection hologram due to interference of the object wave with the incident reference wave, while the other is a transmission hologram formed by the object wave and the totally reflected reference wave.

When the hologram is illuminated with the conjugate of the original reference wave, this is diffracted by the transmission hologram and reconstructs the conjugate of the original object wave. The component not diffracted by this hologram, which is totally reflected at the surface of the emulsion, is then diffracted by the reflection hologram and adds to this reconstructed wave. A real image is formed in the original object plane. Undiffracted light leaves the other face of the prism and does not affect the projected real image.

The proximity of the image plane to the hologram in this setup minimizes

Fig. 12.4. Optical system used to record a hologram with a totally reflected reference wave [Stetson, 1967a].

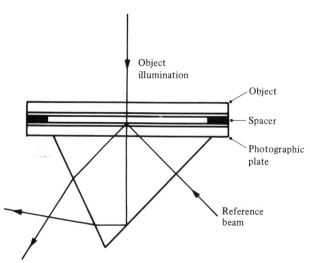

the effects of deviations from flatness of the hologram, and makes the image relatively insensitive to misalignment of the reconstructing beam and the bandwidth of the source. With a proper choice of recording parameters, a resolution of approximately $600 \, \text{mm}^{-1}$ was obtained [Stetson, 1968a].

12.7.   Evanescent-wave holography

These experiments led to the study of holographic techniques using evanescent waves. Evanescent waves occur in diffraction and total reflection [Born & Wolf, 1980], but, because they are exponentially damped, exist only very close to the surface where they are formed. Holography can be used to record evanescent waves; the hologram can then reconstruct the information they carry in the form of homogeneous or propagating waves (see, for example, the review by Bryngdahl [1973]).

As shown schematically in fig. 12.5, an evanescent wave is formed by total reflection at the surface of the photographic emulsion in which the hologram is to be recorded. For this, the emulsion must either be coated on a substrate having a higher refractive index [Nassenstein, 1968b] or immersed in a liquid with a relatively high refractive index [Bryngdahl, 1969b]. The evanescent wave then propagates along the interface. Its wavelength is

$$\lambda_e = \lambda_0/(n_s \sin \theta), \tag{12.7}$$

where λ_0 is the wavelength of the radiation in vacuum, n_s is the refractive index of the medium adjacent to the photographic emulsion and θ is the angle of incidence. We assume that this wave interferes with a homogeneous

Fig. 12.5. Hologram recording with an evanescent wave.

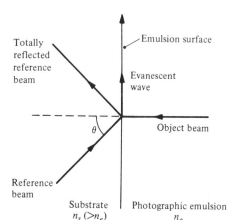

wave incident normally on the interface. The wavelength of this homogeneous wave is

$$\lambda_h = \lambda_0/n_e, \tag{12.8}$$

where n_e is the refractive index of the photographic emulsion.

The complex amplitude of the evanescent wave can be written as

$$a_e = a_r \exp(-2\pi n_s zd) \exp(-i2\pi x/\lambda_e), \tag{12.9}$$

where a_r is the amplitude at the surface, and $d = \lambda_0/(n_s^2 \sin^2\theta - n_e^2)^{1/2}$ is a measure of the depth to which the evanescent wave penetrates into the photographic emulsion. The complex amplitude of the homogeneous wave which may be incident from either side, as shown in figs 12.6(a) and (b), is

$$a_h = a_0 \exp(\pm i2\pi z/\lambda_h). \tag{12.10}$$

The intensity in the interference pattern formed by these two waves is then

$$\begin{aligned} I &= |a_e + a_h|^2, \\ &= a_0^2 + a_r^2 \exp(-4\pi n_s z/d) \\ &\quad + 2a_0 a_r \exp(-2\pi n_s z/d) \cos\{2\pi[(x/\lambda_e) - (z/\lambda_h)]\}. \end{aligned} \tag{12.11}$$

Fig. 12.6. Recording and reconstruction schemes for holograms formed by a homogeneous object wave o and an evanescent reference wave r. Both the recording schemes shown, (a) as well as (b), give the same reconstructed waves with the reconstruction schemes shown in (c), (d), (e) and (f) [Bryngdahl, 1969b].

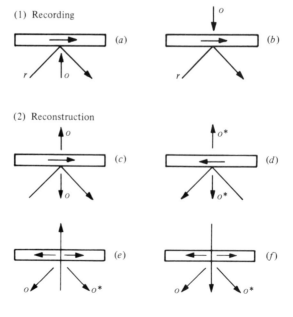

The interference pattern described by the last term in (12.11) is rapidly attenuated as one moves from the surface ($z = 0$) into the emulsion. Typically, the depth of penetration of the evanescent wave is only about 1 μm, so that a thin hologram is formed, with a period equal to the wavelength λ_e of the evanescent wave.

The waves reconstructed by such a hologram differ significantly from those reconstructed by a conventional hologram. Thus, when it is illuminated, as shown in fig. 12.6(c), with the same evanescent wave used to record it, the hologram reconstructs the original homogeneous wave used to record it, and, instead of its conjugate, a mirror-image copy of it propagating in the opposite direction. If, however, the direction of the evanescent wave is reversed, as shown in fig. 12.6(d), the conjugate of the homogeneous wave is reconstructed, along with its mirror image. An interesting feature is that in both of these cases only a single image, unaccompanied by any directly transmitted light, appears on one side of the hologram. Finally, if a homogeneous wave is used to illuminate the hologram, as shown in figs 12.6(e) and 12.6(f), evanescent waves are created which propagate along the hologram. If the plate is immersed in a liquid with a high refractive index, these evanescent waves are converted into homogeneous waves which leave one side of the hologram at angles greater than the critical angle.

It is also possible to record a hologram using two evanescent waves. These can interfere at any angle, provided they have a common component of polarization. Accordingly, if λ_{e1} and λ_{e2} are the wavelengths of the two evanescent waves, the spacing of the hologram fringes can range from

$$\Lambda = \lambda_{e1}\lambda_{e2}/(\lambda_{e1} - \lambda_{e2}), \tag{12.12}$$

when the waves propagate in the same direction, to

$$\Lambda = \lambda_{e1}\lambda_{e2}/(\lambda_{e1} + \lambda_{e2}), \tag{12.13}$$

when they travel in opposite directions. Since from (12.13) the minimum spacing of the fringes is $\lambda_e/2$, and from (12.7) and (12.8), $\lambda_e < \lambda_h$, it is possible to produce finer fringe patterns than with homogeneous standing waves of the same frequency.

In general, an incident evanescent wave is diffracted by such a hologram into other evanescent waves, which, when the boundary conditions are met, are converted into propagating homogeneous waves. An interesting result is that the corresponding wave vectors need not lie in the plane of incidence.

A feature of holograms recorded with evanescent waves is that white light can be used to reconstruct the image [Bryngdahl, 1969b]. This is because, as shown in fig. 12.7, constructive interference occurs for a particular wavelength only when a condition equivalent to the Bragg condition in a volume hologram is satisfied. However, in this case, changes in the thickness of the photographic emulsion do not cause a wavelength shift, because the hologram is recorded only at the surface of the emulsion.

Theoretical studies of the diffraction efficiency of evanescent-wave holograms have been made by Lukosz & Wüthrich [1974], Lee & Streifer [1978a, b] and Woznicki [1980]. These show that if light of the same frequency polarized with the electric vector perpendicular to the plane of incidence (s-polarization) is used for reconstruction as well as for recording, the diffraction efficiency exhibits sharp maxima when either the angle of incidence of the illuminating wave or the angle of diffraction of the reconstructed wave is equal to the critical angle. This is confirmed by experimental measurements [Wüthrich & Lukosz, 1975]. The analysis also shows that the diffraction efficiency for waves polarized with the electric vector in the plane of polarization (p-polarization) is lower and exhibits dips not observed in the case of s-polarization. Quite high diffraction efficiency is possible at the critical angle. This tallies with earlier observations by Nassenstein [1969], who obtained a diffraction efficiency of 0.226 with an amplitude hologram. This is much higher than the theoretical maximum value (0.0625) for an amplitude hologram recorded with homogeneous waves.

An obvious application of evanescent-wave holography is in high-resolution imaging. An advantage here is that the reference and illuminating beams are confined entirely to one side of the recording medium, while the object and the image are on the other side, as shown in figs 12.6(b) and (c). The object can therefore be located very close to the hologram, and an object field with a solid angle close to 2π can be recorded. Nassenstein

Fig. 12.7. Reconstruction by (a) a conventional volume hologram and (b) an evanescent wave hologram. The wavelength selectivity allows white light to be used in both cases [Bryngdahl, 1973].

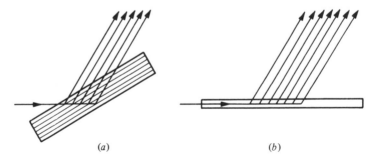

(a)   (b)

[1970] has shown that information about details smaller than the normal resolution limit can be obtained if the object is illuminated with evanescent waves.

Another promising application of holography with evanescent waves is in integrated optics. A hologram recorded in a layer over a planar dielectric waveguide, with a guided mode as the reference wave, can be used as a grating coupler with relatively high efficiency [Lukosz & Wüthrich, 1976; Wüthrich & Lukosz, 1980]. In addition, such holograms open up possibilities of information storage in a form compatible with integrated optics technology.

12.8. Multiple imaging

There are many applications such as, for example, the production of integrated circuits, where it is necessary to produce an array of identical images. While this is normally done by making a series of exposures with a step and repeat camera, it is also possible to use holographic techniques.

12.8.1. Multiple imaging using Fourier holograms

To produce an $n \times m$ array of images separated by intervals x_0, y_0, a hologram is used with an amplitude transmittance

$$H(\xi, \eta) = \sum^n \sum^m \exp\left[-i2\pi(nx_0\xi + my_0\eta)\right]. \tag{12.14}$$

This hologram represents a set of plane waves travelling in directions corresponding to the centres of the images in the array.

This hologram is placed in the back focal plane of the lens L_1 in the optical system shown in fig. 12.8 [Lu, 1968]. If a transparency with an amplitude transmittance $f(x, y)$ located in the front focal plane of L_1 and illuminated by a collimated beam is used as the input, its Fourier transform $F(\xi, \eta)$ is displayed in the back focal plane of L_1.

Accordingly, the wavefront emerging from the hologram can be written as

$$G(\xi, \eta) = F(\xi, \eta)H(\xi, \eta). \tag{12.15}$$

A second Fourier transform operation by the lens L_2 then produces a set of multiple images

$$g(x, y) = f(x, y) * \sum^n \sum^m \delta(x - nx_0, y - my_0),$$

$$= \sum^n \sum^m f(x - nx_0, y - my_0). \tag{12.16}$$

12.8.2. Multiple imaging using lensless Fourier holograms

This technique, developed by Groh [1968], uses a much simpler optical setup. In the first step, a lensless Fourier hologram of an array of point sources $P_1 \dots P_n$ is recorded with a point reference source R.

To use this hologram to generate multiple images, it is illuminated with the conjugate to the original reference wave by means of a lens placed behind it, as shown in fig. 12.9. The hologram then produces real images of the array of object points $P_1 \dots P_n$ in their original positions. If then, the point source is replaced by an illuminated transparency, an array of images of the transparency are formed, centred on the positions of the original point sources $P_1 \dots P_n$.

A variation of these methods [Kalestynski, 1973, 1976] is to record a hologram of the transparency using multiple reference beams. When illuminated with a single reference beam, this hologram then produces an array of images.

Problems arise with all these techniques, due to cross-talk, if the hologram recording is not strictly linear. These can be avoided, at the expense of a considerable reduction in diffraction efficiency, if the hologram is produced by successive exposures using individual object beams separately, rather than all of them together. Another problem is that only the centres of the images are free from aberrations. Accordingly, the individual images must subtend only a small angle at the hologram.

Fig. 12.8. Multiple-image generation by a Fourier hologram [Lu, 1968].

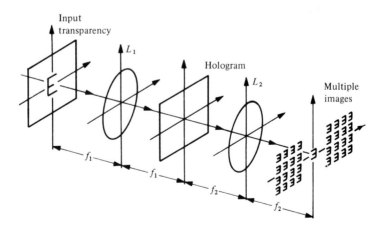

12.9.   Holographic diffraction gratings

Diffraction gratings formed by recording an interference pattern in a suitable light-sensitive medium (now commonly called holographic diffraction gratings) are rapidly replacing conventional ruled gratings in spectroscopy. While Burch & Palmer [1961] first showed that transmission gratings could be made by photographing interference fringes using silver halide emulsions, it was the use of photoresist layers coated on optically worked blanks which finally led to the production of spectrographic gratings of high quality [Rudolph & Schmahl, 1967; Labeyrie & Flamand, 1969]. After processing, these yield a relief image (see section 7.3) which can be coated with an evaporated metal layer and used as a reflection grating.

Holographic gratings have several advantages over ruled gratings. Besides being cheaper and simpler to produce, they are free from periodic and random errors and exhibit much less scattered light. In addition, it is possible to produce much larger gratings of finer pitch, as well as gratings on substrates of varying shapes, and gratings with curved grooves and varying pitch. This makes it possible to produce gratings with unique focusing properties and opens up the possibility of new designs of spectrometers (see, for example, the review by Namioka, Seya & Noda [1976]).

Against this, their main disadvantage is that the groove profile cannot be

Fig. 12.9. Multiple imaging by means of a hologram of an array of point sources [Groh, 1968].

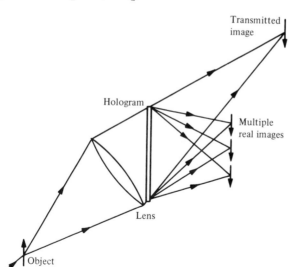

controlled as easily as in ruled gratings. While even a sinusoidal profile can give quite high diffraction efficiencies for very small grating spacings ($\approx \lambda$) [Loewen, Maystre, McPhedran & Wilson, 1975] it is usually necessary to produce a triangular groove profile for maximum diffraction efficiency. Accordingly, a number of methods have been proposed for the production of blazed holographic gratings (see, for example, the reviews by Schmahl & Rudolph [1976] and by Hutley [1976, 1982]).

One method of achieving this result is to expose the photoresist to a sawtooth irradiance distribution, which is built up by a process of Fourier synthesis, either by using more than two beams to produce the fringes, or by making multiple exposures to fringe patterns of appropriate periodicities and phases [McPhedran, Wilson & Waterworth, 1973; Schmahl, 1975; Breidne, Johansson, Nilsson & Åhlen, 1979].

Another possibility which has been explored is to start with a sinusoidal or partially blazed profile and modify it by ion-beam etching to produce a well-formed triangular profile [Aoyagi & Namba, 1976; Aoyagi, Sano & Namba, 1979].

In the most widely used method, however, [Sheridon, 1968; Hutley, 1975] the photoresist layer is aligned obliquely to the fringe pattern, as shown in fig. 12.10. This produces, within the thickness of the resist, layers which are alternately soluble and insoluble. After development, the surface profile is determined by the shape of the insoluble layers near the surface. The only disadvantage of this technique is that one of the beams is incident through the back of the blank, which must therefore be of optical quality.

Fig. 12.10. The production of blazed gratings in photoresist [Hutley, 1982].

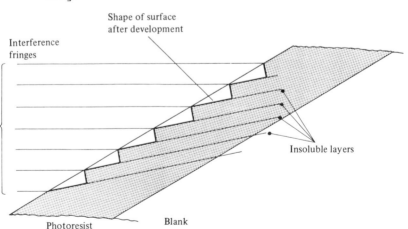

An optical system for this purpose is shown in fig. 12.11. In this, light from an Ar$^+$ laser ($\lambda = 458$ nm) is split into two beams of equal intensity which are focused by microscope objectives on pinholes. Each pinhole is located at the focus of an off-axis parabolic mirror, so that a collimated beam is obtained upon reflection from the mirror. As shown in fig. 12.10, the photoresist-coated blank is placed in the interference field at a small angle to the standing waves.

To produce gratings of good quality, the optics must produce wavefronts plane to $\lambda/10$. The liquid photoresist is applied to the optically worked blank, which is then spun rapidly to produce a uniform layer, about 0.5 μm thick. In addition to the precautions normally taken to ensure stability of the fringes during the exposure, a closed-loop servo system is used to maintain the optical path difference in the interferometer stable to better than $\lambda/50$.

12.10. Holographic scanners

Holographic scanners are a relatively new development which could solve many of the problems associated with mirror scanners. Their most promising applications are in point-of-sale terminals and, with a modulated laser beam, for high-speed non-impact printing.

A simple disc holographic scanner [Cindrich, 1967; McMahon, Franklin & Thaxter, 1969] is shown in fig. 12.12. The disc has a number of holograms

Fig. 12.11. Optical system used to produce blazed holographic gratings [courtesy I. G. Wilson, CSIRO Division of Chemical Physics, Melbourne, Australia].

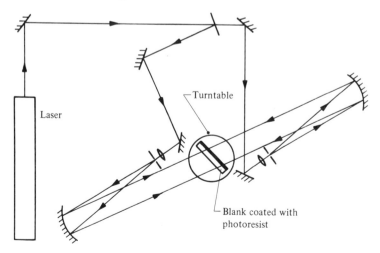

recorded on it, with a point source as the object and a collimated reference beam. Each of these holograms, when illuminated with the conjugate to the reference beam, forms an image of the point source. Rotating the deflector about an axis perpendicular to its surface causes the reconstructed image spot to scan the image plane.

The main problem with this system is that the scanning line is, in general, an arc of a circle with a radius

$$r = d \sin \theta \tan \theta, \tag{12.17}$$

where d is the distance from the scanning facet to the centre of the image plane, and θ is the angle between the diffracted principal ray and the normal to the hologram.

One method of obtaining a straight line scan is to use a cylindrical scanning element such as that shown in fig. 12.13 [Pole & Wolenmann, 1975; Pole, Werlick & Krusche, 1978]. However, such scanners are much more difficult to fabricate than disc scanners, which can be replicated relatively easily.

Another is to use an auxiliary reflector in conjunction with a disc scanner, to make the principal diffracted ray incident normal to the imaging surface [Ih, 1977]. A better arrangement [Kramer, 1981] is shown in fig. 12.14. In this system, a straight line scan is obtained which is self-compensated for wobble of the scanner when $\theta_i = \theta_d = 45°$.

Fig. 12.12. Disc holographic scanner [Kramer, 1981].

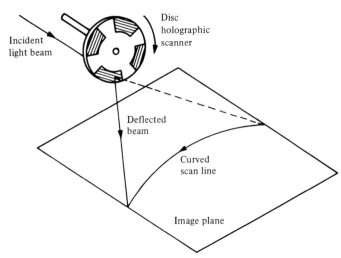

12.11. Holographic optical elements

A hologram can be used to transform an optical wavefront, in much the same manner as a lens [Schwar, Pandya & Weinberg, 1967]. Generalized holographic optical elements (HOEs) have been found very useful in several specialized applications, because they make possible unique system configurations and functions.

One of the main advantages of HOEs is the fact that, unlike conventional optical elements, their function is essentially independent of substrate geometry. In addition, since they can be produced on quite thin substrates, they are relatively light, even for large apertures. Another advantage is the possibility of spatially overlapping elements, since several holograms can be recorded in the same layer. Finally, HOEs provide the possibility of

Fig. 12.13. Cylindrical holographic scanner [Kramer, 1981].

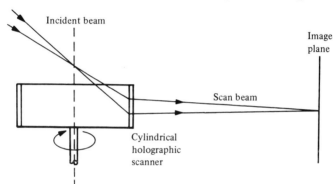

Fig. 12.14. Plane-grating holographic scanner giving a straight-line scan [Kramer, 1981].

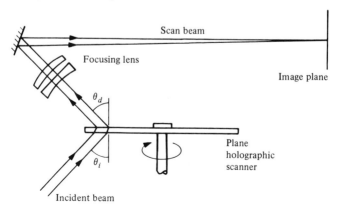

correcting system aberrations, so that separate corrector elements are not required.

A major consideration in the design of systems using HOEs is their optical efficiency, which can vary considerably with the angle and the wavelength. Because of this, a complete analysis of their imaging properties requires a wave-optical treatment [Forshaw, 1973]. This is not very easy in practice, and an approach based on ray tracing is commonly used in system design [Latta, 1971c], calculations being made of the amplitude as well as the direction of the diffracted ray. Analysis of the aberrations (see section 3.4) is made more difficult by the fact that HOEs frequently have to be produced on curved substrates and are used in systems which do not have rotational symmetry. However, conditions for aberration-free imaging have been derived [Welford, 1973; Smith, 1977].

Because the imaging characteristics of HOEs vary considerably with wavelength, they are most commonly used with quasi-monochromatic illumination, though achromatic systems have been proposed by Bennett [1976] and by Sweatt [1977] and studied experimentally by Weingärtner & Rosenbruch [1980a, b, 1982].

Production of HOEs involves recording a hologram in a suitable material using an exposing system which provides the appropriate wavefront [Close, 1975]. However, once the recording system has been set up, HOEs can be produced at a rate limited only by the time taken to prepare, expose and process the substrates. The recording material must have high resolution, good stability, high diffraction efficiency and low scattering. While photoresists, photopolymers and bleached photographic emulsions have been used, dichromated gelatin is, at present, the most widely used material. Electronic fringe stabilization is commonly used to permit the relatively long exposure times required.

Since the recording materials used are sensitive only to blue and violet light, it is necessary to design the system so as to produce, using such a source, a HOE that can give aberration-free performance at the wavelength at which it is to be used [Latta & Pole, 1979]. This can also be done by introducing aberration-compensating elements in the recording setup [Malin & Morrow, 1981].

An attractive possibility is the use of computer-generated holograms to produce the recording wavefronts [Bryngdahl, 1975; Fairchild & Fienup, 1982]. Since computer-generated holograms can produce wavefronts having any arbitrary shape, it is possible to realize a variety of unusual optical components and to achieve geometrical transformations as well as phase transformations.

A simpler method of building up an image light distribution which is acceptable for many applications is by means of a multifacet hologram [Case, Haugen & Løkberg, 1981]. The surface of the hologram is divided into a number of small elements (facets) each of which contains a grating of a specified spatial frequency and orientation. When this hologram is illuminated by a plane or spherical wave, the light incident on a particular facet is diffracted to a specified location in the image plane. The image is thus built up of small patches of light, with near 100 per cent efficiency.

The applications of HOEs are mainly in light-weight systems performing quite complex functions [Mosyakin & Skrotskii, 1972]. These include head-up displays and multiple imaging systems [McCauley, Simpson & Murbach, 1973; Groh, 1968]. They have also found use as beam splitters and beam combiners [Case, 1975] and in multiple-wavelength systems. Since aberration correction is possible via the optical recording setup, elements which perform quite complex phase transformations can be duplicated at relatively low cost.

While HOEs are not likely to replace conventional optical elements, there is little doubt that their use will become increasingly common in a number of specialized applications.

13

Information storage and processing

As mentioned in Chapter 1, the development of holography was greatly stimulated by the application of the concepts of communication theory to optics. Early work on these lines [Elias, Grey & Robinson, 1952; Maréchal & Croce, 1953] led to the development of spatial filtering techniques using coherent light [O'Neill, 1956; Cutrona, 1960; Tsujiuchi, 1963]. In turn, holographic techniques were found extremely useful in the synthesis of complex filters used in image enhancement and restoration, as well as matched spatial filters used for optical pattern recognition [Vander Lugt, 1964].

Optical information processing is now a major field of research (see, for example, Casasent [1978], Lee [1981] for surveys of current techniques and applications). This chapter will therefore be limited to a brief discussion of some aspects having close links with holography as well as the use of holography for information storage.

13.1. Associative storage

A hologram is merely a record of the interference pattern formed by two waves. If the recording process is linear, as defined by (2.2), the transmittance of the hologram can be written as

$$t(x, y) = t_0 + \beta T[|o(x, y)|^2 + |r(x, y)|^2 + o(x, y)r^*(x, y) + o^*(x, y)r(x, y)], \tag{13.1}$$

where t_0 is a uniform background transmittance, T is the exposure time, β is a constant, and $o(x, y)$ and $r(x, y)$ are the complex amplitudes of the object wave and the reference wave, respectively.

Normally, the hologram is illuminated with the reference wave. As shown in Chapter 2, it then reconstructs the object wave. However, it is apparent that if both $|r(x, y)|^2$ and $|o(x, y)|^2$ are constants, (13.1) is symmetrical as far as the two waves are concerned. If, then, the hologram is illuminated with the object wave, it reconstructs the reference wave. Such a situation, where illumination of the hologram with either one of a pair of waves results in

194

reconstruction of the other, can be described by saying that the two complex amplitudes are stored in association.

To obtain the conditions for faithful imaging in such a case, consider the optical system shown in fig. 13.1. This is essentially the same as that used to record and reconstruct a Fourier transform hologram. If two transparencies with amplitude transmittances $f_1(x, y)$ and $f_2(x, y+a)$ located in the front focal plane of the lens L_1 are illuminated by a plane wave of unit amplitude, the resultant complex amplitude in the hologram plane is

$$G(\xi, \eta) = F_1(\xi, \eta) + F_2(\xi, \eta) \exp(-i2\pi\eta a), \tag{13.2}$$

where $F_1(\xi, \eta) \leftrightarrow f_1(x, y)$, and $F_2(\xi, \eta) \leftrightarrow f_2(x, y)$.

Assuming linear recording, the amplitude transmittance of the resulting hologram is

$$\begin{aligned}
t(\xi, \eta) &= t_0 + \beta T |G(\xi, \eta)|^2, \\
&= t_0 + \beta T [|F_1(\xi, \eta)|^2 + |F_2(\xi, \eta)|^2 \\
&\quad + F_1(\xi, \eta) F_2^*(\xi, \eta) \exp(i2\pi\eta a) \\
&\quad + F_1^*(\xi, \eta) F_2(\xi, \eta) \exp(-i2\pi\eta a)].
\end{aligned} \tag{13.3}$$

If this hologram is replaced in the same position in which it was recorded and illuminated by the wave due to $f_2(x, y+a)$ alone, the complex amplitude transmitted by the hologram is

$$\begin{aligned}
H(\xi, \eta) &= t(\xi, \eta) F_2(\xi, \eta) \exp(-i2\pi\eta a), \\
&= t_0 F_2(\xi, \eta) \exp(-i2\pi\eta a) \\
&\quad + \beta T [|F_1(\xi, \eta)|^2 + |F_2(\xi, \eta)|^2] F_2(\xi, \eta) \exp(-i2\pi\eta a) \\
&\quad + \beta T |F_2(\xi, \eta)|^2 F_1(\xi, \eta) \\
&\quad + \beta T F_2^2(\xi, \eta) F_1^*(\xi, \eta) \exp(-i4\pi\eta a).
\end{aligned} \tag{13.4}$$

Fig. 13.1. Optical system used to analyse the conditions for associative storage of two waves.

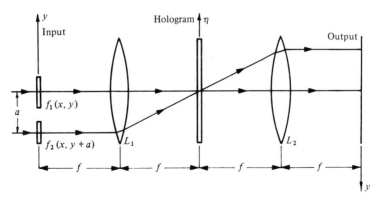

The complex amplitude in the back focal plane of L_2 is then given by the relation

$$h(x, y) \leftrightarrow H(\xi, \eta).$$ (13.5)

The only term of interest in $h(x, y)$ is the primary reconstructed image which is given by the Fourier transform of the third term on the right hand side in (13.4). If we neglect a constant factor, its complex amplitude is

$$h_3(x, y) = \mathscr{F}\{|F_2(\xi, \eta)|^2 F_1(\xi, \eta)\},$$
$$= [f_2(x, y) \star f_2(x, y)] * f_1(x, y).$$ (13.6)

It follows from (13.6) that a perfect image of $f_1(x, y)$ will be reconstructed if the autocorrelation of $f_2(x, y)$ is a delta function. This is obviously the case if $f_2(x, y)$ is a point source.

However, it is also very nearly the situation if $f_2(x, y)$ is a randomly varying transmittance. In this case, the autocorrelation function of $f_2(x, y)$ has a narrow peak superimposed on a relatively weak background whose extent is twice that of $f_2(x, y)$. The finer the spatial structure of $f_2(x, y)$, and the greater its extent, the closer its autocorrelation approximates to a delta function (see Appendix A1.3). In this case, though, it is essential to preserve the geometry of the recording setup for the image to be reconstructed. This follows from (13.6), from which it is apparent that if the reference source moves by the width of the autocorrelation peak, the image will disappear.

The feasibility of using an extended diffusing object as a reference source has been demonstrated in experiments involving what is called 'ghost imaging' [Collier & Pennington, 1966]. Two metal bars, one vertical and the other horizontal, were illuminated with light from a laser, and the scattered light was allowed to fall on a photographic plate. After processing, the photographic plate was replaced precisely in its original position, and the horizontal bar was removed. When the plate was illuminated with the scattered light from the vertical bar, it was found to reconstruct an image of the horizontal bar.

13.2.   Character recognition

This property of associative storage makes it possible to use a hologram for character recognition. This is essentially a spatial filtering operation in which the hologram functions as a matched filter [van Heerden, 1963a].

A typical optical system for this is shown in fig. 13.2. To produce the matched filter, a transparency of the set of characters to be identified is placed in the input plane and a hologram of this transparency is recorded in the Fourier transform plane using a point reference source. For simplicity, we assume that the complex amplitude in the input plane due to the transparency is a one-dimensional distribution

$$f(y) = \sum_{j=1}^{N} f_j(y - c_j), \tag{13.7}$$

where $f_j(y - c_j)$ is the complex amplitude due to a typical character centred at c_j, while that due to the reference source is $\delta(y + b)$.

If we assume linear recording, the transmittance of the hologram can be written as

$$t(\eta) = t_0 + \beta T[1 + |F(\eta)|^2 + F^*(\eta) \exp(-i2\pi\eta b) + F(\eta) \exp(i2\pi\eta b)], \tag{13.8}$$

where $F(\eta) \leftrightarrow f(y)$.

The hologram is replaced, after processing, in exactly the same position in which it was recorded and illuminated by a single character of the set centred on the axis. If the amplitude due to this character in the input plane is $f_l(y)$, the transmitted amplitude at the hologram is

$$\begin{aligned} H(\eta) &= F_l(\eta) t(\eta), \\ &= (t_0 + \beta T) F_l(\eta) + \beta T F_l(\eta) |F(\eta)|^2 \\ &\quad + \beta T F_l(\eta) F^*(\eta) \exp(-i2\pi\eta b) \\ &\quad + \beta T F_l(\eta) F(\eta) \exp(i2\pi\eta b). \end{aligned} \tag{13.9}$$

The complex amplitude in the output plane is then the Fourier transform of (13.9), which is

$$\begin{aligned} h(y) &= (t_0 + \beta T) f_l(y) + \beta T f_l(y) * [f(y) \star f(y)] \\ &\quad + \beta T f_l(y) \star f(y) * \delta(y + b) \\ &\quad + \beta T f_l(y) * f(y) * \delta(y - b). \end{aligned} \tag{13.10}$$

Fig. 13.2. Optical system used for experiments in character recognition.

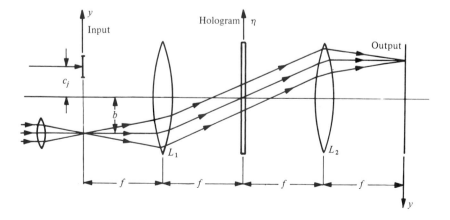

The only term of interest in (13.10) is the last but one on the right hand side which corresponds to the correlation of $f_l(y)$ with all the characters of the set. If we ignore the constant factor βT, this can be expanded as

$$f_l(y)\star f(y)*\delta(y+b)$$

$$=\left[f_l(y)\star \sum_{j=1}^{N} f_j(y-c_j)\right]*\delta(y+b),$$

$$=\left[f_l(y)\star f_l(y)\right]*\delta(y+c_l+b)$$

$$+\left[f_l(y)\star \sum_{\substack{j=1 \\ j\neq l}}^{N} f_j(y-c_j)\right]*\delta(y+b). \tag{13.11}$$

If the autocorrelation function of the character presented is sharply peaked, the first term on the right hand side of (13.11) represents a bright spot of light, which is the reconstructed image of the reference source, located at $y = -c_l - b$. The presence of this bright spot in the output plane corresponds to recognition of the character presented as one belonging to the original set. The fact that this image is reconstructed at a distance $-c_l$ from its correct position identifies the character presented as $f_l(y-c_l)$.

This basic character recognition technique has been extended by Vander Lugt, Rotz & Klooster [1965] to permit simultaneous identification of all the characters on a page.

When real-time operation is not required, a more direct technique can be used [Rau, 1966; Weaver & Goodman, 1966]. This involves the use of two transparencies in the input plane. One of these $f_1(y)$ is a transparency of the character to be located, while the other $f_2(y+b)$ is a transparency of the page of characters to be searched. The transmittance of the Fourier hologram formed with these two sources is then

$$\mathbf{t}(\eta)=\mathbf{t}_0 + \beta T[|F_1(\eta)|^2 + |F_2(\eta)|^2$$
$$+ F_1^*(\eta)F_2(\eta) \exp(-i2\pi\eta b)$$
$$+ F_1(\eta)F_2^*(\eta) \exp(i2\pi\eta b)], \tag{13.12}$$

where $F_1(\eta)\leftrightarrow f_1(y)$ and $F_2(\eta)\leftrightarrow f_2(y)$.

If this hologram is illuminated with a plane wave, the complex amplitude in the output plane is proportional to the Fourier transform of $\mathbf{t}(\eta)$. As before, the only term of interest is the third within the square brackets, which, if we neglect a constant factor, is

$$\mathscr{F}\{F_1^*(\eta)F_2(\eta) \exp(-i2\pi\eta b)\}$$
$$=[f_1(y)\star f_2(y)]*\delta(y+b). \tag{13.13}$$

If $f_2(y)$ is identical with $f_1(y)$, this term will result in a bright autocorrelation peak at $y = -b$. If, however, $f_2(y)$ contains more than one such character $f_j(y - c_j)$, identical with $f_1(y)$ but located at different positions, an equal number of autocorrelation peaks will be formed at locations $y = -b - c_j$, corresponding to the centres of these patterns.

13.3. Coding and multiplexing

The technique of associated storage described in section 13.2 need not be restricted to the case of a single pair of wavefronts. It is possible to record holograms of a series of subjects on the same plate using a different coded reference wave for each hologram. Typically, this can be done by using an illuminated ground-glass screen as the reference source and moving the screen through a small distance ($\approx 20\ \mu$m) between exposures. If the processed hologram is replaced in exactly the same position in which it was recorded, any one of the stored images can be recovered separately by illuminating the multiplexed hologram with the appropriate coded reference wave, as can be seen from the following analysis [La Macchia & White, 1968].

Let $O_i(\eta)$ be the complex amplitude due to the ith object at the hologram and $R_i(\eta)$ be that due to the corresponding coded reference wave. If N such exposures are recorded on the same photographic plate, the transmittance of the hologram is

$$t(\eta) = t_0 + \beta T \sum_{i=1}^{N} |O_i(\eta) + R_i(\eta)|^2. \tag{13.14}$$

If the processed hologram is replaced in exactly the same position in which it was recorded and illuminated by the jth reference wave, the transmitted amplitude is

$$H(\eta) = R_j(\eta)t(\eta). \tag{13.15}$$

The only term in the expansion of (13.15) which is of interest is, as before, that corresponding to the primary reconstructed images. This is, apart from a constant factor

$$
\begin{aligned}
H_3(\eta) &= R_j(\eta) \sum_{i=1}^{N} R_i^*(\eta)O_i(\eta), \\
&= R_j(\eta)R_j^*(\eta)O_j(\eta) \\
&\quad + R_j(\eta) \sum_{\substack{i=1 \\ i \neq j}}^{N} R_i^*(\eta)O_i(\eta).
\end{aligned}
\tag{13.16}
$$

The complex amplitude in the image plane is then

$$h_3(y) = \mathscr{F}\{H_3(\eta)\},$$
$$= r_j(y) \star r_j(y) * o_j(y)$$
$$+ r_j(y) \star \sum_{\substack{i=1 \\ i \neq j}}^{N} r_i(y) * o_i(y). \tag{13.17}$$

The first term on the right hand side of (13.17) corresponds to the reconstructed image of the *j*th object (the signal). This is superposed on a diffuse halo due to the remaining cross-correlation terms (noise).

Since the diffracted power corresponding to each of the terms in the expansion of (13.17) is the same, it follows that the ratio of the signal power to the total noise power, measured in the hologram plane, is

$$I_S/I_N = 1/N. \tag{13.18}$$

This could result in a very poor signal-to-noise ratio (see section 6.8) when the number of holograms multiplexed on the same plate is large, but for the fact that we are concerned with the ratio of the intensities in the image plane. The most favourable case is where the image is that of a point source; the image power is then concentrated in a small, relatively bright spot, while the noise power is spread out over an area having twice the lateral extent of the coded source, and hence of relatively low intensity. In such a case, it is possible to superimpose more than 1000 coded reference holograms of individual points on a single plate, the practical limit being set mainly by the drop in the diffraction efficiency of the individual holograms, which is inversely proportional to the square of the number of exposures (see section 4.10).

13.4.   Image processing

Holographic spatial filtering can also be used to improve a picture taken with an imperfectly corrected optical system, provided its impulse response is known [Stroke & Zech, 1967]. We assume that the picture is a positive transparency, which has been processed so that the product of the values of γ (see Appendix A5) for the negative and positive materials is equal to 2. The amplitude transmittance of the transparency can then be written as

$$h(x, y) = f(x, y) * g(x, y), \tag{13.19}$$

where $f(x, y)$ would have been its amplitude transmittance with a perfectly corrected optical system, and $g(x, y)$ is the impulse response of the system. The latter can be defined by another transparency, which is the image of a bright point recorded, under the same conditions, with the same optical system.

To retrieve $f(x, y)$, the transparency $h(x, y)$ is placed in the front focal plane of the lens L_1, in an optical system such as that shown in fig. 13.1 and illuminated with a collimated beam. The complex amplitude in the back focal plane of L_1 is then the Fourier transform of $h(x, y)$ and is given by the relation

$$H(\xi, \eta) = F(\xi, \eta)G(\xi, \eta), \tag{13.20}$$

where $F(\xi, \eta) \leftrightarrow f(x, y)$ and $G(\xi, \eta) \leftrightarrow g(x, y)$. This can be rewritten as

$$F(\xi, \eta) = H(\xi, \eta)/G(\xi, \eta),$$
$$= H(\xi, \eta)[G^*(\xi, \eta)/|G(\xi, \eta)|^2]. \tag{13.21}$$

It is apparent from (13.21) that if a filter with an amplitude transmittance $G^*(\xi, \eta)/|G(\xi, \eta)|^2$ is inserted in the back focal plane of L_1, the complex amplitude in the back focal plane of L_2 will yield $f(x, y)$. Such a filter can be produced by superposing two filters with amplitude transmittances of $|G(\xi, \eta)|^{-2}$ and $G^*(\xi, \eta)$.

To produce the first filter, the transparency $g(x, y)$ is placed in the front focal plane of L_1, and the resulting intensity distribution in its back focal plane is recorded on a photographic plate. If this plate is developed with $\gamma = 2$, its amplitude transmittance becomes

$$t_1(\xi, \eta) = [|G(\xi, \eta)|^2]^{-\gamma/2},$$
$$= |G(\xi, \eta)|^{-2}. \tag{13.22}$$

To produce the second filter, the optical system is modified, as shown in fig. 13.2, by the introduction of a collimated reference beam, and a Fourier hologram of $g(x, y)$ is recorded on a fresh photographic plate placed in the back focal plane of L_1. Assuming linear recording, as defined by (2.2), the transmittance of this hologram can be written as

$$t_2(\xi, \eta) = t_0 + \beta T[1 + |G(\xi, \eta)|^2 + G(\xi, \eta)\exp(i2\pi\eta b)$$
$$+ G^*(\xi, \eta)\exp(-i2\pi\eta b)]. \tag{13.23}$$

The last term on the right hand side of (13.23) contains the required transmittance. Hence, when the two filters $t_1(\xi, \eta)$ and $t_2(\xi, \eta)$ are superposed and placed in the back focal plane of L_1, and the transparency $h(x, y)$ is illuminated with a plane wave, the corresponding term in the output plane yields $f(x, y)$.

Simplified techniques by which both the filters can be recorded on the same plate have been described by Zetsche [1982] and by Jo & Lee [1982].

13.5. Space-variant operations

A space-variant, linear operation can be defined as one whose effect on a point in the input field depends on the location of this point. Such operations are of considerable interest in data processing; their basic properties and some of the optical techniques which can be used to perform them have been reviewed by Walkup [1980], Goodman [1981] and Rhodes [1981].

To carry out a two-dimensional linear space-variant operation it is necessary to have a system whose impulse response is a function of four independent variables (two more than a normal optical system). Two methods based on holographic techniques will be described in this section.

The first is a simple method developed by Bryngdahl [1974*a*, *b*] to perform the coordinate transformation

$$x = G_1(u, v), \tag{13.24}$$

$$y = G_2(u, v). \tag{13.25}$$

This transformation is effected with the optical system shown in fig. 13.3, which uses a computer-generated hologram whose spatial frequencies at any point (u, v) are

$$s_u = G_1(u, v)/\lambda f \tag{13.26}$$

and

$$s_v = G_2(u, v)/\lambda f. \tag{13.27}$$

Light from a point in the input plane having coordinates (u, v) is then diffracted at an angle such that an image of this point is formed in the back focal plane of the lens L_2 at a point whose coordinates (x, y) satisfy (13.24) and (13.25).

More general operations can be realized, in principle, by a hologram array. Each input pixel is backed by a hologram element which generates

Fig. 13.3. Optical system used to produce a coordinate transformation.

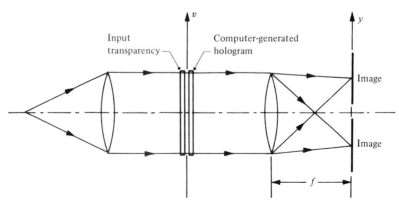

the desired impulse response for that pixel. However, there are serious limitations on the number of pixels which can be handled in this fashion, due to the limited resolution of the hologram elements when they are made very small.

Another method of obtaining a space-variant impulse response is to use a thick holographic element as a filter in the spatial frequency plane in an optical system such as that shown in fig. 13.2 [Deen, Walkup & Hagler, 1975]. This filter contains a number of superimposed holograms, each recorded with a plane reference wave incident at a different angle. Each point in the input plane gives rise to a plane wave whose angle of incidence on the filter depends on the coordinates of this point and, hence, generates an impulse response determined by the corresponding hologram. However, to avoid cross-talk, the input field must contain only a small number of input points, since all points lying on a cone satisfy the Bragg condition.

Higher selectivity can be obtained by the use of coded reference beams [Krile, Marks, Walkup & Hagler, 1977; Jones, Walkup & Hagler, 1982]. For this, a diffuser is inserted in the input plane, and each of the holograms in the filter is recorded with a reference beam derived from a small area on this diffuser. Since the autocorrelation functions of the reference beams are sharply peaked, each point in the input produces an output from the corresponding filter. The diffuse background arising from the cross-correlation functions of the elementary diffusers is minimized by using a thick recording medium.

13.6.   Information storage

Information can be stored in a compact form in microfilm. However, the maximum useful storage density with microfilm is set by the fact that, beyond a certain point, dust or scratches can result in total loss of significant parts of the stored information. Holographic information storage has the advantage (see section 3.7) that surface damage does not wipe out any particular item of information, but only results in a drop in the overall signal-to-noise ratio. This makes it possible to use materials with much higher resolution and achieve much higher storage densities.

13.6.1.   Holographic memories

The virtual immunity of information stored on a hologram to degradation is particularly valuable where storage of information in a binary code for a computer is involved. This led at an early stage to the construction of page-organized holographic memories [Anderson, 1968; Langdon, 1970]. As shown schematically in fig. 13.4, these used a 32×32

array of small holograms, each containing 1024 bits of information. Two acousto-optic cells in tandem allowed x and y deflection of the laser beam so that any one of these holograms could be addressed. When illuminated by the laser beam, the information stored in the hologram was read out by a detector array in the image plane. This gave a capacity of about 10^6 bits with an access time of a few microseconds.

These systems were initially developed as read-only stores. Subsequently, Stewart *et al.* [1973] showed that with a suitable recording medium which could be recycled, such as a photothermoplastic, and a suitable page-composer, such a system could also be used as a read–write memory, though with a relatively long cycle time.

However, semiconductor memories are now available with comparable capacities and with much shorter access times. Since there is not much scope for reducing the access time of holographic memories, work over the past few years has mainly been in the direction of increasing their capacity, and, thereby, decreasing the storage cost per bit [Kiemle, 1974; Knight, 1975*b*].

The capacity of a simple holographic memory, such as that described earlier, is limited mainly by the resolution of the auxiliary optics to about 10^8 bits [Vander Lugt, 1973]. One approach to increased capacity is to use a number of memory modules, each consisting of a holographic storage plate with its own detector array, which are accessed, as shown in fig. 13.5, by a common laser and beam deflector [Lang & Eschler, 1974]. Such a system can provide a storage capacity of approximately 2×10^{10} bits and, in principle, can operate in a read–write configuration as well, using either a liquid-crystal page-composer or a multifrequency acousto-optic page-composer [Eschler, 1975].

An alternative approach has been to use the increased storage capacity

Fig. 13.4. Optical system for a read-only, page-organized, holographic information store [Kogelnik, 1972].

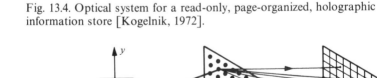
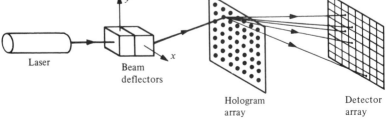

Laser

Beam
deflectors

Hologram
array

Detector
array

made available by recording several holograms in the same thick recording medium [van Heerden, 1963b; Friesem & Walker, 1970]. Selective readout is achieved by means of electro-optic deflectors and beam translators which permit random access as well as Bragg-angle selection [D'Auria, Huignard, Slezak & Spitz, 1974]. Such a system can provide capacities of 10^{11}–10^{12} bits.

A major obstacle which must be overcome before such large holographic memories can become competitive with conventional erasable memories for general computer applications is the development of a better storage medium. Iron-doped $LiNbO_3$ has a limited dynamic range and recording and erasure are relatively slow. An alternative which has been proposed is $Sr_{0.75}Ba_{0.25}Nb_2O_6$ [Thaxter & Kestigian, 1974].

One of the most interesting lines of development as well as, potentially, the most important is the construction of a parallel-search holographic memory [Gabor, 1969]. This makes use of the associative properties of holographic storage and would be extremely useful in applications involving rapid retrieval of data from a very large data base. In such a system [Knight, 1974, 1975a; Gerasimova & Zakharchenko, 1981] illumination of the memory plane with a search code produces an image in the detector plane for each hologram which contains a record of a logical match for the search code. This information is then used to steer the reference beam to each of these holograms, in turn, to read out the data stored in it.

13.6.2. Specialized information stores

While holographic memories for digital computers have still to fulfil their initial promise, holographic information storage has been much more successful in meeting various specialized needs. One of these is in reducing the storage space required for archival copies of documents below

Fig. 13.5. Schematic of a modular holographic memory [Lang & Eschler, 1974].

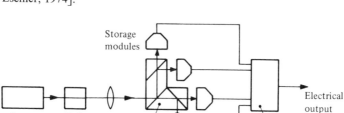

what is possible with microfilms. This is done by recording holograms (1–2 mm diameter) of each microfilm frame [Vagin, Nazarova, Arseneva & Vanin, 1975]. The extent to which high-density storage and faithful grey-tone reproduction can be obtained with such systems has been studied by Killat [1977*b*]. Multicolour material such as maps and motion pictures can be stored using holograms which have much better archival stability than colour film [Gale, Knop & Russell, 1975; Ih, 1975; Gale & Knop, 1976; Yu, Tai & Chen, 1978].

Holograms have also been used to provide additional machine-readable information on a conventional microfiche [Nelson, Vander Lugt & Zech, 1974]. An exploratory holographic information–processing system for libraries has been described by Tsukamoto *et al.* [1974] in which abstracts of literature are stored in holographic arrays on 35 mm roll film. Relevant abstracts can be located by interrogating the system with the appropriate keywords, using holographic correlation, after which the data is read out using a laser deflector in the same way as in a conventional holographic memory. A large-capacity, high-speed, holographic system for filing patents has been described by Sugaya, Ishikawa, Hoshino & Iwamoto [1981]. This can store up to 280 000 pages any one of which can be retrieved within 1 second.

Another interesting application of holography has been in videotape and videodisk systems [Hannan, Flory, Lurie & Ryan, 1973; Tsunoda, Tatsumo & Kataoka, 1976]. A simple colour-encoding technique for this purpose has been proposed by Nishihara & Koyama [1979]. Digitized audio messages have been recorded as small one-dimensional Fourier-transform holograms on a disk in an audio response system with a 2000 word vocabulary and 250 M bit/s transfer rate [Kubota *et al.*, 1980]. A holographic digital record/reproduce system has been described using a multichannel acousto-optic modulator array and a mode-locked cavity-dumped laser to achieve a data rate of 500 M bit/s [Roberts, Watkins & Johnson, 1974].

However, the most interesting possibilities for commercial exploitation of holographic information storage appear to be in credit- and identity-card verification. Systems for this purpose have been described by Sutherlin, Lauer & Olenick [1974], Abramson, Bjelkhagen & Skande [1979], and Greenaway [1980].

14

Holographic interferometry

Holographic interferometry is an extension of interferometric measurement techniques in which at least one of the waves which interfere is reconstructed by a hologram.

The unique advantages of holographic interferometry arise from the fact that holography permits storing a wavefront for reconstruction at a later time. Wavefronts which were originally separated in time or space, or even wavefronts of different wavelengths, can be compared by holographic interferometry. As a result, changes in the shape of objects with quite rough surfaces can be studied with interferometric precision.

One of the most important applications of holographic interferometry is in non-destructive testing (see for example, Erf [1974] and Vest [1981]). It can be used wherever the presence of a structural weakness results in localized deformation of the surface when the specimen is stressed, either by the application of a load or by a change in pressure or temperature. Crack detection and the location of areas of poor bonding in composite structures are fields where holographic interferometry has been found very useful. An allied area of applications has been in medical and dental research, where it has been used to study the deformations of anatomical structures under stress, as well as for nondestructive tests on prostheses. Some of these are reviewed in a paper by Greguss [1976] as well as in two volumes edited by Greguss [1975] and by von Bally [1979].

Holographic interferometry has also proved its utility in aerodynamics, heat transfer, and plasma diagnostics. Yet another field of application has been in solid mechanics, where it has been used to evaluate the stresses in complex structures, as well as to measure changes in shape due to corrosion or absorption of water. Since mechanical contacts are not involved, measurements can be carried out even in hostile or corrosive environments. Many of these applications have been discussed in detail by Vest [1979].

14.1. Real-time holographic interferometry

In this technique, the hologram is replaced, after processing, in exactly the same position in which it was recorded. When it is illuminated

207

with the original reference beam, the virtual image coincides with the object. If, however, the shape of the object changes very slightly, two sets of light waves reach the observer, one being the reconstructed wave (corresponding to the object before the change) and the other the directly transmitted wave from the object in its present state. The two wave amplitudes will add at the points where the difference in optical paths is zero or a whole number of wavelengths, and cancel at some other points in between. As a result, an observer viewing the reconstructed image sees it covered with a pattern of interference fringes, which is a contour map of the changes in shape of the object. These changes can be observed in real time.

If we consider a typical off-axis holographic recording system, as in section 2.2, the intensity at the photographic plate when the hologram is recorded is

$$I(x, y) = |r(x, y) + o(x, y)|^2, \tag{14.1}$$

where $r(x, y)$ is the complex amplitude due to the reference beam and $o(x, y) = |o(x, y)| \exp[-i\phi(x, y)]$ is the complex amplitude due to the object in its normal state.

Assuming, as in (2.2), that the amplitude transmittance of the photographic plate after processing is linearly related to the exposure, the amplitude transmittance of the hologram is

$$\mathbf{t}(x, y) = \mathbf{t}_0 + \beta T I(x, y), \tag{14.2}$$

where, as before, β is the slope (negative) of the amplitude transmittance *versus* exposure characteristic of the photographic material, T is the exposure time and $\mathbf{t}_0$ is a uniform background transmittance.

When the processed hologram is replaced in the same position in which it was recorded, it is illuminated by the wave from the deformed object as well as the reference wave. Accordingly, the complex amplitude of the wave transmitted by the hologram is

$$u(x, y) = [o'(x, y) + r(x, y)]\mathbf{t}(x, y), \tag{14.3}$$

where $o'(x, y)$ is the complex amplitude of the wave from the deformed object. If the change in the shape of the object is very small, it can be assumed that only the phase distribution of the object wave is modified, so that

$$o'(x, y) = |o(x, y)| \exp[-i\phi'(x, y)]. \tag{14.4}$$

The only terms of interest in the expansion of (14.3) are those corresponding to the primary image and the directly transmitted wave. If $|r(x, y)|^2 = r^2$, the complex amplitude due to these is

$$u'(x, y) = \beta T r^2 o(x, y) + (\mathbf{t}_0 + \beta T r^2) o'(x, y). \tag{14.5}$$

Accordingly, the resultant intensity is

$$I'(x, y) = |o(x, y)|^2 \{\beta^2 T^2 r^4 + (t_0 + \beta T r^2)^2$$
$$+ 2\beta T r^2 (t_0 + \beta T r^2) \cos[\phi'(x, y) - \phi(x, y)]\}. \qquad (14.6)$$

The reconstructed image is covered with fringes. Since β is negative, a dark fringe corresponds to the condition

$$\phi'(x, y) - \phi(x, y) = 2n\pi, \qquad (14.7)$$

where n is an integer.

The visibility of the interference pattern is a maximum when

$$|\beta T r^2| = |t_0 + \beta T r^2|, \qquad (14.8)$$

which, since β is negative, corresponds to the condition

$$|\beta T r^2| = t_0/2. \qquad (14.9)$$

Precise repositioning of the hologram after processing is necessary to avoid the introduction of spurious fringes. In addition, with photographic emulsions, it is necessary to take precautions to avoid local deformations of the emulsion due to nonuniform drying. These problems can be avoided by exposing and processing the photographic plate *in situ* in a liquid gate [van Deelen & Nisenson, 1969]. A typical arrangement is shown in fig. 14.1. In this setup, processing is speeded up by using a monobath (see section 7.1.4);

Fig. 14.1. Setup for real-time holographic interferometry. The hologram is processed *in situ*, and the interference fringes are viewed through a closed-circuit television system.

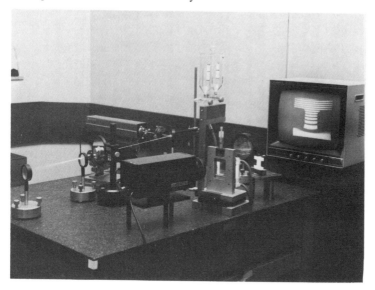

this also makes it possible to use a simple gravity system for handling the processing solutions [Hariharan & Ramprasad, 1973b]. A closed-circuit television camera is used to project the fringe pattern on the screen of a black-and-white television monitor. The fringes can be colour-coded to identify the sign of the displacement by photographing them with a Polaroid camera through red and green filters with a small change in the load between the exposures [Hariharan, 1977a].

An alternative which completely eliminates the need for wet processing is to use thermoplastic recording [Thinh & Tanaka, 1973]. Automated systems using such materials are now commercially available with which a hologram can be recorded and viewed in less than a minute. Since these produce a thin phase hologram, the diffraction efficiency is high, and effects due to changes in the thickness of the recording material are eliminated. A possibility, where maximum dimensional stability of the hologram is required, is to use solvent-vapour processing [Saito, Imamura, Honda & Tsujiuchi, 1980]. It is worth noting that with such a phase hologram, the reconstructed image exhibits a phase shift of $\pi/2$ [see Hariharan & Hegedus, 1975] so that a dark zero-order fringe is no longer obtained.

A problem in real-time holographic interferometry is that while the light diffracted by the hologram remains linearly polarized, the light scattered by a diffusely reflecting object is largely depolarized (see section 5.3). This results in a serious drop in the visibility of the fringes. To avoid this, it is necessary to use a polarizer when viewing or photographing the fringes.

14.2.   Double-exposure holographic interferometry

In double-exposure holographic interferometry, interference takes place between the wavefronts reconstructed by two holograms of the object recorded on the same photographic plate. Typically, the first exposure is made with the object in its initial, unstressed condition, and the second is made with a stress applied to the object. When the processed hologram is illuminated with the original reference beam, it reconstructs two images, one corresponding to the object in its unstressed state, and the other corresponding to the stressed object. The resultant interference pattern reveals the changes in shape of the object between the two exposures.

In this case, the intensity at the photographic plate during the first exposure is

$$I_1(x, y) = |r(x, y) + o(x, y)|^2, \qquad (14.10)$$

and that during the second exposure is

$$I_2(x, y) = |r(x, y) + o'(x, y)|^2. \qquad (14.11)$$

The amplitude transmittance of the resulting hologram is, therefore,

$$\mathbf{t}(x, y) = \mathbf{t}_0 + \beta T(I_1 + I_2). \tag{14.12}$$

When the hologram is illuminated once again with the same reference wave, the transmitted amplitude in the hologram plane is

$$u(x, y) = r(x, y)\mathbf{t}(x, y). \tag{14.13}$$

The only terms of interest in the expansion of (14.13) are those corresponding to the two superimposed primary images. The complex amplitude due to these is

$$u_3(x, y) = \beta T r^2 |o(x, y)|\{\exp[-i\phi(x, y)] \\ + \exp[-i\phi'(x, y)]\}, \tag{14.14}$$

so that the resultant intensity is

$$I_3(x, y) \propto |o(x, y)|^2 \{1 + \cos[\phi(x, y) - \phi'(x, y)]\}. \tag{14.15}$$

In this case the hologram is a permanent record of the change in shape of the object.

Double-exposure holographic interferometry is much easier than real-time holographic interferometry, because the two interfering waves are always reconstructed in exact register. Distortions of the emulsion affect both images equally, and no special care need be taken in illuminating the hologram when viewing the image. In addition, since the two diffracted wavefronts are similarly polarized and have almost the same amplitude, the fringes have very good visibility.

However, double-exposure holographic interferometry also has certain limitations. The first is that where the object has not moved between the exposures, the reconstructed waves, both of which have experienced the same phase shift, add to give a bright image of the object. This makes it difficult to observe small displacements. A dark field, which gives much higher sensitivity, can be obtained by holographic subtraction. For this the phase of the reference beam is shifted by π between the two exposures [Collins, 1968; Hariharan & Ramprasad, 1972].

An alternative, which also helps to resolve ambiguities in the sense of the displacement, is to shift the phase of the reference beam by $\pi/2$ between the two exposures, or, better, to introduce a very small tilt in the wavefront illuminating the object between the two exposures. This results in reference fringes whose position is modulated by the phase shifts being studied [Jahoda, Jeffries & Sawyer, 1967; Hariharan & Ramprasad, 1973a].

Another limitation is that information on intermediate states of the object is lost. This can be overcome to some extent by multiplexing techniques using spatial division of the hologram [Caulfield, 1972;

Hariharan & Hegedus, 1973]. In the latter, a series of masks are used in which the apertures overlap in a systematic fashion, and a sequence of holograms is recorded at different stages of loading. The images can then be reconstructed, two at a time, so that interference patterns between any two images can be studied. An alternative is a method described by Parker [1978], using a thermoplastic recording material, by which real-time fringes can be observed and the fringe pattern then frozen to give a permanent holographic record.

An interesting possibility which has been studied by Yu & Chen [1978] is the use of rainbow holograms for double-exposure holographic interferometry. Different views of the fringes and fringe patterns due to different effects can be displayed in different colours [Tai, Yu & Chen, 1979; Yu, Tai & Chen, 1979].

14.3.   Sandwich holograms

Control of the fringes, to compensate for rigid body motion and eliminate ambiguities in interpretation, is not normally possible with a doubly-exposed hologram. However, it is possible with two holograms recorded with two, different, angularly separated reference waves [Gates, 1968; Ballard, 1968; Tsuruta, Shiotake & Itoh, 1968]. These may be either on the same plate or on different plates.

An elegant alternative is the sandwich hologram [Abramson, 1974, 1975, 1977; Abramson & Bjelkhagen, 1979]. In this technique, as shown in fig. 14.2, pairs of photographic plates (without any antihalation backing) are exposed in the same plate holder with their emulsion-coated surfaces facing the object. B_1, F_1 are exposed with the unstressed object, while B_2, F_2 and $B_3, F_3, \ldots$, are exposed with the object deformed by increasing loads. After all the plates have been processed, B_1 is combined with, say, F_2 in the original plateholder and illuminated with the original reference beam, to produce an interference pattern showing the surface deformations at the corresponding stage of loading, as shown in fig. 14.3(a). A tilt of the sandwich then results in a change in the interference pattern exactly equivalent to a tilt of the original object as shown in fig. 14.3(b). A detailed theoretical analysis of the changes in the fringes has been made by Dubas & Schumann [1977].

If B_1 is combined with $F_2, F_3, \ldots$, it is possible to study the total deformation at any stage, while combinations such as $B_1F_2, B_2F_3, B_3F_4, \ldots$ will show the incremental deformations. A simplified version of this technique is also possible using only two plates cemented together with a spacer [Hariharan & Hegedus, 1976].

Fig. 14.2. Steps involved in sandwich hologram interferometry [Abramson, 1975].

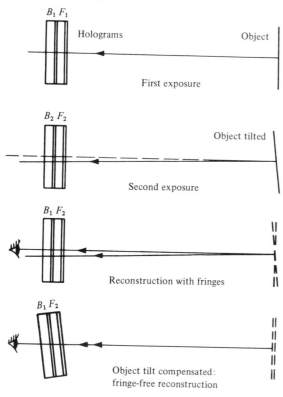

Fig. 14.3. Interference fringes obtained with a sandwich hologram using a thin, circular metal sheet clamped at its edge and subjected to a bending moment about the horizontal diameter, (*a*) when the sandwich is replaced in its original position and (*b*) when reference fringes are introduced by tilting the sandwich about a vertical axis [Hariharan & Hegedus, 1976].

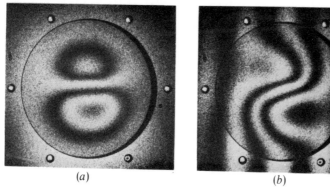

(*a*)   (*b*)

14.4.   Holographic interferometry in an industrial environment

Holographic interferometry normally requires an extremely stable optical setup. However, various techniques have been described which permit double-exposure holographic interferometry in an industrial environment.

One method is the use of a pulsed laser. Double-exposure holographic interferometry can then be used to study transient phenomena such as deformations due to impact loading as shown in fig. 14.4 [Gates, Hall & Ross, 1972; Armstrong & Forman, 1977]. An electrical, optical or acoustic signal is used to trigger the first pulse just before, or at the instant of impact, with the second pulse following after a predetermined delay. Techniques such as sandwich holography can be used to eliminate unwanted rigid body displacements and simplify interpretation of the fringes [Bjelkhagen, 1977b; Abramson & Bjelkhagen, 1978].

Even objects rotating at extremely high speeds can be studied with an optical derotator, [Stetson, 1978] consisting of an inverting prism mounted in the hollow shaft of an electric motor. When this prism is aligned axially with the rotating object and rotates at half its speed, a stationary image is obtained. The residual movement is small enough to permit recording double-pulse holographic interferograms of quite large rotating objects (see fig. 14.5).

A technique which permits holographic interferometry under some

Fig. 14.4. Holographic interferogram recorded with a double-pulsed laser showing the effect of a hammer blow on a crash helmet. Pulse duration 25 ns, pulse separation 25 μs [Gates, Hall & Ross, 1972; Crown copyright, National Physical Laboratory, reprinted with permission].

conditions with a continuous-wave laser source is object motion compensation. This is possible when the unwanted movements of the object are mainly out-of-plane translation or vibration, and is achieved by reflecting the reference wave also from a suitably chosen point on the surface, or from a mirror attached to it [Mottier, 1969; Waters, 1972]. Another method is to use reflection ('piggyback') holography (see section 5.1). Two holograms are recorded by illuminating the object through the hologram plate. Relative motion is virtually eliminated, since the hologram plate moves with the object [Neumann & Penn, 1972; Boone, 1975].

Image holography can also be used to minimize the effects of object movement, particularly when studying out-of-plane deformations [Klimenko, Matinyan & Dubitskii, 1975; Rowley, 1979, 1981].

14.5. Holographic interferometry of phase objects

Even in applications such as flow visualization and heat transfer studies, where conventional interferometry has been used for many years, holographic interferometry has practical advantages (see, for example, Tanner [1966] and Trolinger [1975b]).

Fig. 14.5. Holographic interferogram of a vibrating turbine fan recorded while rotating at 4460 rpm using an image derotator and a double-pulsed laser [courtesy K. A. Stetson, United Technology Research Center, East Hartford, USA].

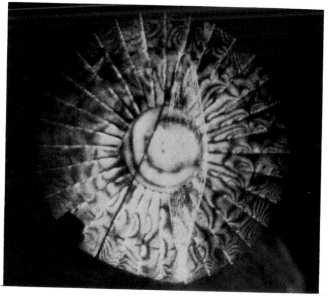

In the first instance, mirrors and windows of relatively low optical quality can be used. Since the phase errors due to these contribute equally to both interfering wavefronts, they cancel out, and only the effects of changes in the optical path are seen. However, the most significant advantage is the possibility of incorporating a diffusing screen (a ground glass plate) in the interferometer. This gives an interference pattern that is localized near the phase object, and can be viewed and photographed over a range of angles. This makes it possible to study three-dimensional refractive index distributions.

If the refractive index gradients in the test section are assumed to be small, so that rays propagate through it along straight lines parallel to the z axis, $\phi(x, y)$ the phase difference at any point in the interference pattern is given by the relation

$$\phi(x, y) = k_0 \int [n(x, y, z) - n_0] dz, \tag{14.16}$$

where n_0 is the refractive index of the medium in the test section in its initial, unperturbed state and $n(x, y, z)$ is the final refractive index distribution.

The simplest case is that of a two-dimensional phase object with no variation of refractive index in the z direction. In this case the refractive index distribution can be calculated directly from (14.16). Fortunately, this is a valid approximation to many practical situations.

Another case which lends itself to analytic treatment is that of a refractive index distribution $f(r)$ which is radially symmetric about an axis normal to the line of sight (for convenience, say, the y axis).

For a ray travelling in the z direction at a distance x from the centre, we then have

$$dz = (r^2 - x^2)^{-1/2} r dr, \tag{14.17}$$

so that (14.16) becomes

$$\phi(x, y) = 2 \int_x^\infty f(r)(r^2 - x^2)^{-1/2} r dr. \tag{14.18}$$

This is the Abel transform of $f(r)$, and it can be inverted to find $f(r)$ [Bracewell, 1965].

The evaluation of an asymmetric refractive index distribution $f(r, \theta)$ is much more difficult and is possible only by recording a large number of interferograms from different directions [Sweeney & Vest, 1973]. The problem becomes even more complicated when the effects of ray curvature due to refraction cannot be neglected. An iterative technique which can be used under these conditions has been described by Cha & Vest [1979, 1981].

Holographic interferometry has been found particularly useful in plasma diagnostics [Zaidel, Ostrovskaya & Ostrovskii, 1969]. Since, unlike a neutral gas, a plasma is highly dispersive, measurements of the refractive index distribution at two wavelengths make it possible to determine the electron density directly. For this, two holograms are recorded simultaneously on the same plate with light from a ruby laser which has passed through a frequency doubler to produce two collinear beams with wavelengths $\lambda_1 = 694$ nm and $\lambda_2 = 347$ nm. If the plate is processed to give a nonlinear recording characteristic (see section 6.5) it is possible to make the second-order image reconstructed by one hologram interfere with the first-order image reconstructed by the other. Under these conditions it can be shown [Ostrovskaya & Ostrovskii, 1971] that the interference fringes are contours of constant dispersion and, hence, of constant electron density. Measurements carried out by this technique on a plasma produced in air by the beam from a CO_2 laser have been described by Radley [1975].

14.6. Specialized techniques in holographic interferometry of phase objects

Many of the techniques of conventional interferometry such as multiple-beam and multiple-pass interferometry as well as shearing interferometry can be extended to holographic interferometry [see Vest, 1979]. A type of shear possible only with holography is longitudinally reversed shear (phase-conjugate interferometry): for this, the primary and conjugate images of the test wavefront reconstructed by a hologram are made to interfere [Bryngdahl, 1969a; Fainman, Lenz & Shamir, 1981]. This technique can be used to obtain considerably increased sensitivity [Matsuda, Freund & Hariharan, 1981]. Phase-difference amplification is also possible using the higher diffracted orders from a nonlinear hologram [Bryngdahl & Lohmann, 1968a; Matsumoto & Takashima, 1970].

Interesting possibilities for phase-conjugate interferometry have been opened up [Hopf, 1980] by developments in dynamic holography which make it possible to generate a conjugate wave in real time. Optical systems for phase-conjugate interferometry have been described using degenerate four-wave mixing in a thin film of eosin [Bar-Joseph, Hardy, Katzir & Silberberg, 1981] and in a BGO crystal slice [Ja, 1982].

14.7. Interference pattern with a diffusely reflecting object

One of the major advantages of holographic interferometry is that it is possible to obtain interference patterns of high visibility even with an object having a relatively rough surface. To understand how this is possible, and to obtain a relation between the interference order in the fringe pattern and the surface displacement at the corresponding point on the object, consider a small area of the surface which has suffered a simple translation as shown in fig. 14.6, so that two points on it, P and Q, undergo identical vector displacements $\mathbf{L}$ to P' and Q' respectively.

Let the complex amplitude of the wave reflected by the undeformed object be

$$o_1(x, y) = |o(x, y)| \exp[-i\phi(x, y)], \tag{14.19}$$

where $\phi(x, y)$ varies in a random manner over the surface because of its microstructure. The complex amplitude of the wave reflected by the deformed object can then be written as

$$o_2(x, y) = |o(x, y)| \exp\{-i[\phi(x, y) + \Delta\phi]\}, \tag{14.20}$$

where $\Delta\phi$ is constant over the small area considered.

The intensity due to these two wavefronts is then

$$\begin{aligned}
I(x, y) &= |o_1(x, y) + o_2(x, y)|^2, \\
&= o_1(x, y)o_1^*(x, y) + o_2(x, y)o_2^*(x, y) \\
&\quad + o_1(x, y)o_2^*(x, y) + o_1^*(x, y)o_2(x, y). \tag{14.21}
\end{aligned}$$

The resulting interference pattern exhibits a granular structure due to the rapid variation of $\phi(x, y)$ over the surface. However, if this is ignored, the average intensity over a small area centred on (x, y) containing a number of speckles is

$$\begin{aligned}
I &= \langle|o_1 + o_2|^2\rangle, \\
&= \langle o_1 o_1^* \rangle + \langle o_2 o_2^* \rangle \\
&\quad + \langle o_1 o_2^* \rangle + \langle o_1^* o_2 \rangle,
\end{aligned}$$

Fig. 14.6. Fringe formation in holographic interferometry with a diffusely reflecting object.

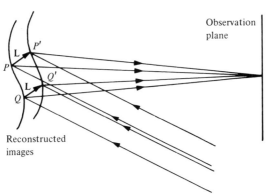

$$= I_1 + I_2 + (I_1 I_2)^{1/2} \exp{(i\Delta\phi)} + (I_1 I_2)^{1/2} \exp{(-i\Delta\phi)},$$
$$= I_1 + I_2 + 2(I_1 I_2)^{1/2} \cos{\Delta\phi}, \tag{14.22}$$

where I_1 and I_2 are the average intensities due to the individual wavefronts.

It is therefore enough to consider the path differences between corresponding points on the two wavefronts to evaluate the broader variations in intensity which constitute the fringes in the interference pattern.

The phase difference $\Delta\phi$ between the two wavefronts, which gives rise to the fringes seen by the observer, can now be found from the change in the total optical path from the source S to the observer O. For small displacements ($|\mathbf{L}| \ll SP, PO$) as shown in fig. 14.7, $\Delta\phi$ is given by the relation

$$\Delta\phi = \mathbf{L} \cdot (\mathbf{k}_1 - \mathbf{k}_2),$$
$$= \mathbf{L} \cdot \mathbf{K} \tag{14.23}$$

where $\mathbf{k}_1$ and $\mathbf{k}_2$ are the propagation vectors of the incident and scattered light. These are of magnitude $|\mathbf{k}_1| = |\mathbf{k}_2| = k_0 = 2\pi/\lambda$ and are taken along the directions of illumination and observation respectively [Aleksandrov & Bonch-Bruevich, 1967; Ennos, 1968; Sollid, 1969]. The sensitivity vector $\mathbf{K} = \mathbf{k}_1 - \mathbf{k}_2$ defined by (14.23) points along the bisector of the angle 2ψ between the illumination and viewing directions, and its magnitude is $|\mathbf{K}| = 2k_0 \cos\psi$.

14.8. Fringe localization

The visibility of the interference fringes formed with a diffusely reflecting object, as described in section 14.7, is a maximum for a particular position of the plane of observation. This position is commonly termed the plane of localization of the fringes. An early finding [Haines & Hildebrand, 1966] was that the position of the plane of localization depends on the type

Fig. 14.7. Calculation of the phase difference in the interference pattern.

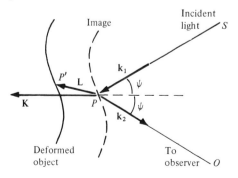

of displacement undergone by the object. Difficulties arise when the fringes are localized too far from the object for both the fringes and the object to be imaged sharply at the same time.

Fringe localization is due to the random phase variations across the object wavefront and is linked with the finite aperture of the viewing system. Detailed analyses of the phenomenon have been made by a number of authors including Stetson [1969, 1970*b*, 1974], Walles [1969], Steel [1970] and Dubas & Schumann [1974]. These showed that, actually, the fringes are not localized in a plane but along a line. Detailed experiments have also been carried out by Molin & Stetson [1970*a, b*, 1971] on the localization of the fringes for different types of object displacements.

However, while fringe localization played a major part in early methods of interpretation of holographic interferograms, it can now be seen as a separate issue. Accordingly, this section is confined to a simple treatment of the phenomenon and a discussion of two limiting cases.

As shown in section 14.7, only waves from corresponding points on the two wavefronts effectively contribute to the fringes. Hence, we need only consider the phase differences between a pair of such points P and P', on the interfering wavefronts, over a range of viewing directions defined by the aperture of the viewing system, as shown in fig. 14.8. At an arbitrarily chosen plane, this phase difference $\Delta\phi$ will, in general, vary over this cone due to the changes in the sensitivity vector in (14.23). However, a plane can be found at a distance from the object determined by the geometry of the system and the magnitude and direction of the displacement at which the value of $\Delta\phi$ is very nearly constant over this range of viewing directions.

Fig. 14.8. Localization of fringes in holographic interferometry with a diffusely reflecting object.

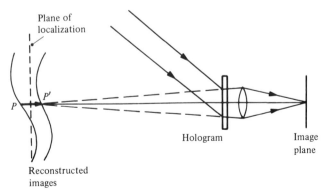

Plane of
localization

P'

P

Hologram

Image
plane

Reconstructed
images

Accordingly, if the imaging system is focused on this plane, a fringe system of good visibility will be seen.

14.8.1. Pure translation of the surface

For pure translation, as shown in fig. 14.9, L is constant over the whole object. If we assume that the object is illuminated with a plane wave, so that $\mathbf{k}_1$ is a constant, the phase difference $\Delta\phi$ is

$$\begin{aligned} \Delta\phi &= \mathbf{L}\cdot(\mathbf{k}_1-\mathbf{k}_2), \\ &= \text{a constant}-\mathbf{L}\cdot\mathbf{k}_2. \end{aligned} \tag{14.24}$$

The condition for localization ($\Delta\phi=$ constant) is then simply

$$\mathbf{k}_2 = \text{a constant}, \tag{14.25}$$

over the detector, implying that the fringes are localized at infinity.

If a lens is placed at a distance equal to its focal length from the surface, plane waves corresponding to different values of $\mathbf{k}_2$ are brought to a focus at different points in its back focal plane. Fringes are then seen in this plane due to the corresponding variations of $(\mathbf{k}_1-\mathbf{k}_2)$. Since the variation of $(\mathbf{k}_1-\mathbf{k}_2)$ is a maximum in the plane containing $\mathbf{k}_1$ and $\mathbf{k}_2$ and is negligible at right angles to it, these are parallel straight fringes.

14.8.2. Pure rotation about an axis in the surface

Unlike the case of pure translation, pure rotation of the object about an axis contained in its surface results in straight-line fringes localized very close to the surface.

As shown in fig. 14.10, the surface is considered to be initially in the xy plane and to rotate about the y axis through a small angle θ. It is also assumed that the surface is illuminated by a collimated beam and that the directions of illumination and observation lie in the xz plane and make angles ψ_1 and ψ_2, respectively, with the z axis. The phase difference between

Fig. 14.9. Fringe localization for pure translation of the surface.

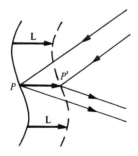

the waves scattered from a pair of corresponding points P and P' is then, from (14.23),

$$\phi = -(2\pi/\lambda)x\theta(\cos\psi_1 + \cos\psi_2). \qquad (14.26)$$

This corresponds to a set of straight fringes running parallel to the y axis with a spacing equal to $\lambda/\theta\ (\cos\psi_1 + \cos\psi_2)$.

To find the surface of localization of the fringes, we consider the variations of this phase difference over a narrow range of viewing directions centred on an arbitrary point $O(x_0, z_0)$ along the line of sight. For this, (14.26) is differentiated to give the relation

$$d\phi = -(2\pi/\lambda)[\theta(\cos\psi_1 + \cos\psi_2)dx - \theta x \sin\psi_2 d\psi_2], \qquad (14.27)$$

since ψ_1 does not vary. Now,

$$\tan\psi_2 = (x - x_0)/z_0, \qquad (14.28)$$

so that

$$dx = (z_0/\cos^2\psi_2)d\psi_2. \qquad (14.29)$$

Hence,

$$d\phi = -(2\pi/\lambda)\theta[(z_0/\cos^2\psi_2)(\cos\psi_1 + \cos\psi_2) - x\sin\psi_2]d\psi_2. \qquad (14.30)$$

For the fringes to be localized at O, $(d\phi/d\psi_2)$ must be equal to zero, so that

$$z_0 = (x\sin\psi_2\cos^2\psi_2)/(\cos\psi_1 + \cos\psi_2). \qquad (14.31)$$

If the viewing direction is normal to the surface of the object, so that $\psi_2 = 0$, then $z_0 = 0$, and the fringes are localized on the surface of the object.

Fig. 14.10. Fringe localization for pure rotation about an axis in the surface.

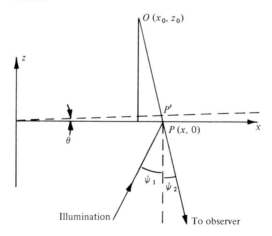

If $\psi_2 \neq 0$, the surface of localization lies either in front of the object (when $\psi_2 > 0$) or behind it (when $\psi_2 < 0$). However, since $(\mathrm{d}\phi/\mathrm{d}\psi_2) = 0$ when $x = 0$, the surface of localization always intersects the surface of the object at the axis of rotation.

14.9. The holodiagram

The holodiagram [Abramson, 1969, 1970a, b, 1971, 1972] is a device which provides a simple geometrical representation of the relations discussed earlier. For the basic hologram recording system shown in fig. 14.11, the holodiagram consists, as shown in fig. 14.12, of a set of spheroids with their foci at O and O'. Each of these spheroids is the locus of points for which the distance OPO' is a constant, and this distance changes in steps of λ from one spheroid to the next. For simplicity, it is usually enough to consider the ellipses formed by the intersection of these spheroids with the xy plane.

A shift of one fringe in the interference pattern is caused by the movement of P from one ellipse to the next. The displacement of P is a minimum when its motion is normal to the ellipse, that is to say, along the sensitivity vector **K**. The holodiagram also shows that while a displacement of P of $\lambda/2$ along the normal to the ellipse, when P lies on the x axis, results in a shift of one

Fig. 14.11. Schematic of a holographic system.

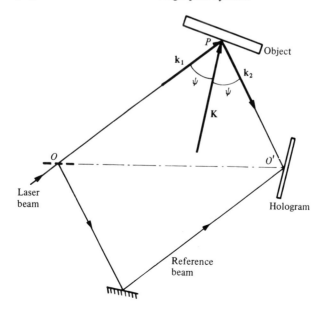

fringe, at any other location a larger displacement $q\lambda/2$, where $q = 1/\cos\psi$, normal to the ellipse is required for the same fringe shift. The corresponding variations in the magnitude of the sensitivity vector are conveniently represented on the holodiagram by curves of constant q. These are the loci of points for which the angle $OPO' = 2\psi$ is a constant, and are, therefore, as shown in fig. 14.12, circles passing through O and O'. The centres of these circles are located on the y axis at points given by the relation

$$y = (l/2)(2-q^2)(q^2-1)^{-1/2}, \tag{14.32}$$

where $OO' = 2l$. The value of q obtained from the holodiagram permits direct evaluation of the fringe pattern. If the fringe order at a point P on the object is N, the component of the displacement normal to the ellipse passing through P is merely $qN\lambda/2$.

The holodiagram is particularly valuable as an aid to an understanding of the fringe patterns which are obtained for different types of object displacements when the distances OP and PO' are comparable to OO', since the sensitivity vector then varies considerably over the image field. It is also

Fig. 14.12. The holodiagram showing (a) ellipses of constant path length, and (b) circles along which the sensitivity vector is a constant [Abramson, 1969].

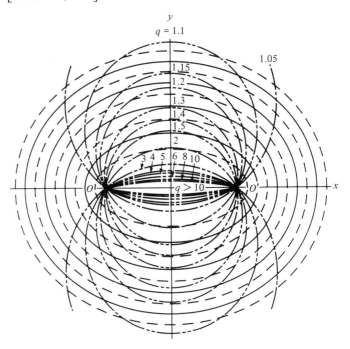

very useful to optimize a hologram recording setup. Since the system is most sensitive to motion normal to the ellipses and least sensitive to tangential motion, the holodiagram can be used to design the system to maximize or minimize its sensitivity to a particular type of object motion. It can also be used when holograms of a relatively large object are to be recorded with a laser of limited coherence length Δl. For this, a holodiagram is drawn in which the spacing of the ellipses corresponds to increments in the path difference of Δl. If, then, the mirror which reflects light for the reference beam is placed on one ellipse, the visible portions of the object must lie within the area bounded by the adjacent two ellipses.

14.10. Holographic strain analysis

Inspection of the fringe pattern gives a considerable amount of information on the type and magnitude of surface movements. This is quite useful to detect areas of stress concentration or localized defects. However, quantitative stress analysis requires measurement of the strains.

If at any point on the stressed object, L_x, L_y and L_z denote the x, y and z components, respectively, of the displacement, the three components of normal strain at this point are defined by the relations

$$\varepsilon_x = \partial L_x / \partial x, \tag{14.33}$$

$$\varepsilon_y = \partial L_y / \partial y, \tag{14.34}$$

$$\varepsilon_z = \partial L_z / \partial z, \tag{14.35}$$

while the three shear strains are

$$\gamma_{xy} = (\partial L_x / \partial y) + (\partial L_y / \partial x), \tag{14.36}$$

$$\gamma_{yz} = (\partial L_y / \partial z) + (\partial L_z / \partial y), \tag{14.37}$$

$$\gamma_{zx} = (\partial L_z / \partial x) + (\partial L_x / \partial z). \tag{14.38}$$

Several methods of analysis of the fringe pattern have been proposed to evaluate the surface displacements and, hence, the strains, and most of these have been discussed in an exhaustive review by Briers [1976].

Early workers tended to favour methods using observations of fringe localization. These have the advantage (see Dubas & Schumann [1974, 1975]) that they can give direct measurements of the surface displacement and, in some cases, even permit direct evaluation of the strain [Stetson, 1976]. However, the accuracy is limited, and interpretation of the fringes becomes difficult when a combination of rotations and translations is involved.

Methods of differentiating the displacement data to obtain the local strain, making use of the moire patterns obtained by superposing two transparencies of the interferogram, have also been described [Boone & Verbiest, 1969; Stetson, 1970a]. These methods can give a quick picture of the surface strains but, again, are of limited applicability.

Another direct method is the fringe vector method [Stetson, 1974, 1975a, b, 1979]. This is based on the fact that any combination of homogeneous deformation and rotation of an object yields fringes that appear to be produced by the intersection of the object surface with a number of equally spaced surfaces which are contours of constant phase difference (the fringe locus function). The fringe vector $\mathbf{K}_f$ runs perpendicular to these surfaces and its magnitude is inversely proportional to their separation.

If we consider a three-dimensional object such as that shown in fig. 14.13, the fringe vector must be normal to the plane defined by the points 1, 2, 3 which lie on the same fringe. Hence, $\hat{\mathbf{k}}_f$ the unit vector in the direction of the fringe vector is given by the relation

$$\hat{\mathbf{k}}_f = (\mathbf{R}_{12} \times \mathbf{R}_{13})/|\mathbf{R}_{12} \times \mathbf{R}_{13}|, \tag{14.39}$$

while the magnitude of the fringe vector is

$$|\mathbf{K}_f| = 2\pi/\hat{\mathbf{k}}_f \cdot \mathbf{R}_{14}. \tag{14.40}$$

Accordingly, if the shape of the object is known, the fringe vector can be evaluated.

To apply this method, the fringe vectors $\mathbf{K}_{f1}, \mathbf{K}_{f2}, \mathbf{K}_{f3}$, corresponding to three different directions of viewing, and, hence, to three different values $\mathbf{K}_1, \mathbf{K}_2, \mathbf{K}_3$, of the sensitivity vector, are determined. Since it can be shown

Fig. 14.13. Fringes due to homogeneous deformation of a three-dimensional object. The points 1, 2 and 3 define the plane of a fringe surface, while the point 4 is on the adjacent fringe [Stetson, 1975b].

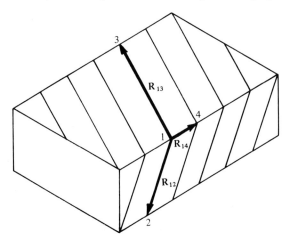

that the resolved components of the fringe vectors and the sensitivity vectors along the x, y and z axes are linked to the gradients of the displacements by the matrix relation

$$\begin{bmatrix} K_{f1x} & K_{f1y} & K_{f1z} \\ K_{f2x} & K_{f2y} & K_{f2z} \\ K_{f3x} & K_{f3y} & K_{f3z} \end{bmatrix} = \begin{bmatrix} K_{1x} & K_{1y} & K_{1z} \\ K_{2x} & K_{2y} & K_{2z} \\ K_{3x} & K_{3y} & K_{3z} \end{bmatrix} \begin{bmatrix} \partial L_x/\partial x & \partial L_y/\partial x & \partial L_z/\partial x \\ \partial L_x/\partial y & \partial L_y/\partial y & \partial L_z/\partial y \\ \partial L_x/\partial z & \partial L_y/\partial z & \partial L_z/\partial z \end{bmatrix}$$

(14.41)

the strains can be evaluated using (14.33) to (14.38).

This method can be extended to more general deformations, which can often be treated as approximately homogeneous over a limited region. Corrections can also be introduced for the variations in the sensitivity vector due to perspective over the region of the object which is being studied [Pryputniewicz & Stetson, 1976]. Experimental studies using this method have been described by Pryputniewicz & Bowley [1978], Pryputniewicz [1978, 1980] and Pryputniewicz & Stetson [1980].

The other approach which has been followed to calculate the strains involves evaluating and differentiating the actual surface displacements. The process is simplified by the fact that, in most cases, what is involved is not the absolute displacement of any point, but rather its displacement with respect to a point in the field of view which can be assumed to be stationary.

The latter can be evaluated if three observations of the fringe order are made with three different directions of observation [Shibayama & Uchiyama, 1971]. However, three separate holograms are then required to give adequate angular separation, and a better alternative is to use a single direction of observation and three different directions of object illumination [Hung, Hu, Henley & Taylor, 1973]. The measured phase differences $\Delta\phi_1$, $\Delta\phi_2$, $\Delta\phi_3$, are then linked to L_x, L_y, L_z, the three orthogonal components of the displacement by the matrix relation

$$\begin{bmatrix} K_{1x} & K_{1y} & K_{1z} \\ K_{2x} & K_{2y} & K_{2z} \\ K_{3x} & K_{3y} & K_{3z} \end{bmatrix} \begin{bmatrix} L_x \\ L_y \\ L_z \end{bmatrix} = \begin{bmatrix} \Delta\phi_1 \\ \Delta\phi_2 \\ \Delta\phi_3 \end{bmatrix} .$$

(14.42)

Data reduction can be simplified by illuminating the object from four different directions making equal angles with the viewing direction, two in the same vertical plane and two in the same horizontal plane [Goldberg, 1975]. This also provides a check on the accuracy of the measurements.

The experimental errors in evaluating **L** from (14.42) fall into two categories: those related to the geometry of the measuring system and those associated with the measurement of the optical phase. The former have been analysed by Matsumoto, Iwata & Nagata [1973] and by Nobis & Vest [1978] and can be minimized by making the angular separation of the sensitivity vectors as large as possible.

The accuracy with which the optical phase can be determined from measurements on photographs is limited by the fact that it is often difficult to locate the centres of the fringes to better than 0.1 of their spacing. In addition, when the number of fringes is small and they are unequally spaced, ambiguities can arise and further errors are introduced by the need for non-linear interpolation to determine fractional fringe orders. The same problem also arises in numerical differentiation of the data when measurements are not available at closely-spaced, regular intervals. In either case, it is necessary to fit a polynomial to the data which can then be used for interpolation or differentiated analytically.

While higher accuracy can be obtained by using television techniques to store and process the fringes to find their centre lines [Kreis & Kreitlow, 1980; Schluter, 1980; Nakadate, Magome, Honda & Tsujiuchi, 1981], a better way out of these problems is the use of electronic techniques which permit direct measurements of the optical phase difference $\Delta\phi$ at a uniformly spaced network of points.

14.11.   Heterodyne holographic interferometry

In this technique [Dändliker, Ineichen & Mottier, 1973] two holograms are recorded of the object, using the setup shown in fig. 14.14. The two holograms are recorded on the same plate with two angularly separated reference beams having the same frequency as the object beam. However, at the reconstruction stage, a small frequency shift is introduced between the frequencies of the two beams illuminating the hologram. This can be done by means of a rotating grating or by means of two acousto-optic modulators operated at slightly different frequencies. The electric fields corresponding to the two reconstructed waves can then be written as

$$E_1(x, y, t) = |u_1(x, y)| \exp\{i[2\pi\nu_1 t - \phi_1(x, y)]\} \tag{14.43}$$

and

$$E_2(x, y, t) = |u_2(x, y)| \exp\{i[2\pi\nu_2 t - \phi_2(x, y)]\}, \tag{14.44}$$

where ν_1, ν_2 are the frequencies of the two reference beams, $|u_1(x, y)|$ and $|u_2(x, y)|$ are the real parts of the amplitudes, and $\phi_1(x, y)$, $\phi_2(x, y)$ are the phases of the two reconstructed waves.

When these two waves are superposed, the intensity at a photodetector is

$$I(x, y, t) = |E_1(x, y, t) + E_2(x, y, t)|^2,$$
$$= |u_1|^2 + |u_2|^2 + 2|u_1||u_2|\cos[2\pi(v_1 - v_2)t - \Delta\phi], \quad (14.45)$$

where $\Delta\phi = \phi_1(x, y) - \phi_2(x, y)$. The power at the detector is modulated at the beat frequency $(v_1 - v_2)$ and the phase of this modulation corresponds exactly to the optical phase difference between the interfering beams.

Since the beat frequency can be resolved by the photodetector, the modulation can be separated by a filter whose passband is centred on this frequency, and its phase $\Delta\phi$ can be measured electronically with respect to a reference signal at the same frequency. This can be derived from a second, fixed detector, while the first detector is moved in steps across the interference field.

This technique can measure the phase with an accuracy of $2\pi/500$ and has been used for a variety of measurements [Dändliker, 1980].

14.12. Digital holographic interferometry

An alternative to heterodyne holographic interferometry which is more suitable for rapid measurements on real-time fringes at a large number of points is digital holographic interferometry [Hariharan, Oreb & Brown, 1982; Dändliker, Thalmann & ·Willemin, 1982].

In this technique, as shown in fig. 14.15, an image of the real-time interference fringes formed by the object and the image reconstructed by the hologram is picked up by a solid-state television camera. The light sensitive

Fig. 14.14. Experimental setup for heterodyne holographic interferometry [Dändliker, 1980].

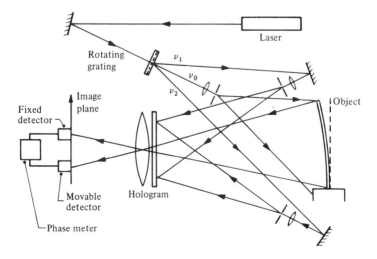

element in this camera is a 100×100 array of photodiodes formed on a single chip. The output voltage from this device at any instant corresponds to the intensity at the particular diode element which is being addressed.

During a single scan of the array, which normally takes about 20 ms, values of the intensity at this 100×100 array of points covering the interference pattern are stored in the memory of a computer. Between successive scans of the array, the phase of the reference beam is shifted relative to that of the object beam, first by $+2\pi/3$ and then by $-2\pi/3$, by means of a mirror mounted on a piezoelectric translator to which

Fig. 14.15. System for digital holographic interferometry [Hariharan, Oreb & Brown, 1982].

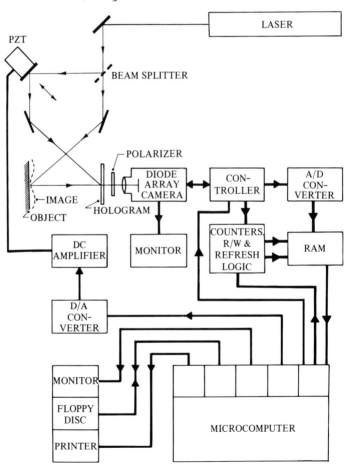

appropriate voltages are applied by a dc amplifier controlled through a digital-to-analog converter by the computer.

At the end of a measurement cycle consisting of three successive scans, the memory of the computer contains three readings of the intensity at each of the 100 × 100 points in the interference pattern imaged on the array. If the original phase difference between the interfering wavefronts at any point on the object is $\Delta\phi$, these readings correspond to phase differences of $\Delta\phi$, $\Delta\phi + 2\pi/3$ and $\Delta\phi - 2\pi/3$, so that the corresponding intensities at the point (x, y) can be written as

$$I_1 = I_R + I_O + 2(I_R I_O)^{1/2} \cos \Delta\phi, \tag{14.46}$$

$$I_2 = I_R + I_O - (I_R I_O)^{1/2} (\cos \Delta\phi + \sqrt{3} \sin \Delta\phi), \tag{14.47}$$

and

$$I_3 = I_R + I_O - (I_R I_O)^{1/2} (\cos \Delta\phi - \sqrt{3} \sin \Delta\phi), \tag{14.48}$$

where I_R is the intensity of the reconstructed image and I_O is that of the object at the same point. From (14.46), (14.47) and (14.48) it follows that

$$3^{-1/2} \tan \Delta\phi = (I_3 - I_2)/(2I_1 - I_2 - I_3). \tag{14.49}$$

The right hand side of (14.49) is evaluated quite simply by the computer. Values of $\Delta\phi$ can then be obtained with a nominal accuracy of $2\pi/200$ from a look-up table.

This phase-measurement system has been used [Hariharan, Oreb & Brown, 1983] in conjunction with an optical system which permits four holograms to be recorded in quick succession on a photothermoplastic material with the object illuminated from four different directions. Phase data from these holograms are then processed in the microcomputer to give the components of the vector displacement at each of the 100 × 100 points.

15

Holographic interferometry: further applications

15.1. Holographic interferometry of vibrating surfaces

Holographic interferometry can also be used to map the amplitude of vibration of a diffusely reflecting surface. The most commonly used technique for this purpose is called time-average holographic interferometry.

15.1.1. Time-average holographic interferometry

In this technique a hologram is recorded of the vibrating surface with an exposure time which is long compared to the period of vibration [Powell & Stetson, 1965].

Consider a point $P(x, y)$ on the object whose displacement at time t is given by the relation

$$\mathbf{L}(x, y, t) = \mathbf{L}(x, y) \sin \omega t. \tag{15.1}$$

The phase shift of the light scattered from this point is then a function of time which, from (14.23), can be written as

$$\Delta\phi(x, y, t) = \mathbf{K} \cdot \mathbf{L}(x, y) \sin \omega t, \tag{15.2}$$

where $\mathbf{K}$ is the sensitivity vector.

Now, let $o(x, y) = |o(x, y)| \exp\left[-i\phi(x, y)\right]$ represent the complex amplitude of the scattered light from P when the object is stationary in its equilibrium position. The complex amplitude of the scattered light from the vibrating object at any instant is then

$$o(x, y, t) = |o(x, y)| \exp\{-i[\phi(x, y) + \mathbf{K} \cdot \mathbf{L}(x, y) \sin \omega t]\}. \tag{15.3}$$

Since the phase of the object wave changes very slowly with time compared to the electric field the holographic recording process can be thought of, in this case, as involving the recording of a very large number of superposed holograms, one for each slightly displaced position of the object. Accordingly, if the holographic recording process is assumed to be linear, the complex amplitude $u(x, y)$ of the wave reconstructed by the hologram will be proportional to the time average of $o(x, y, t)$ over the exposure interval T, so that we can write

$$u(x, y) = \frac{1}{T} \int_0^T |o(x, y)| \exp\{-i[\phi(x, y) + \mathbf{K} \cdot \mathbf{L}(x, y) \sin \omega t]\} \mathrm{d}t,$$

$$= |o(x, y)| \exp[-i\phi(x, y)] \frac{1}{T} \int_0^T \exp[-i\mathbf{K} \cdot \mathbf{L}(x, y) \sin \omega t] \mathrm{d}t,$$

$$= o(x, y) M_T(x, y), \tag{15.4}$$

where $M_T(x, y)$ is known as the characteristic function. If the exposure time is long compared to the period of vibration ($T \gg 2\pi/\omega$), we have

$$M_T(x, y) = \lim_{T \to \infty} \frac{1}{T} \int_0^T \exp[-i\mathbf{K} \cdot \mathbf{L}(x, y) \sin \omega t] \mathrm{d}t,$$

$$= J_0[\mathbf{K} \cdot \mathbf{L}(x, y)], \tag{15.5}$$

where J_0 is the zero-order Bessel function of the first kind. The intensity in the reconstructed image is then

$$I(x, y) = |o(x, y) M_T(x, y)|^2,$$

$$= I_0(x, y) J_0^2[\mathbf{K} \cdot \mathbf{L}(x, y)], \tag{15.6}$$

where $I_0(x, y)$ is the intensity when the object is at rest. The function $|M_T|^2$ is plotted against the parameter $\Omega = \mathbf{K} \cdot \mathbf{L}$ in fig. 15.1(a). If the vibration amplitude varies across the object, (15.6) gives rise to fringes (contours of equal vibration amplitude) covering the reconstructed image. The dark fringes, at which the intensity drops to zero, correspond to the zeros of the function $J_0^2(\Omega)$, and the bright fringes to its maxima. The first maximum,

Fig. 15.1. Characteristic functions for a sinusoidally vibrating object corresponding to (a) time-average fringes and (b) real-time fringes.

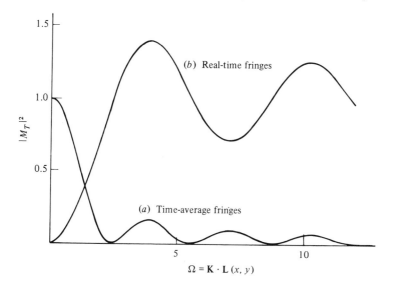

which corresponds to the nodes, is the brightest, while successive maxima occurring at larger vibration amplitudes fall off progressively.

Time-average holography permits ready identification of the vibration modes as well as accurate measurements of the vibration amplitude in many cases. A typical series of interferograms is presented in fig. 15.2. These show the time-average fringes obtained with a model of an aircraft tail-fin excited at different frequencies corresponding to some of its principal vibration modes.

Apart from the fact that the fringes can be seen only after processing the hologram, the main limitation of this technique is that it does not give any information on the relative phases of the vibration at different points on the object. In addition, it cannot be used for very small vibration amplitudes ($|\mathbf{L}| < \lambda/8$).

15.1.2. Real-time interferometry of vibrating surfaces

Real-time holographic interferometry (see section 14.1) is useful to identify the resonances of a test object by monitoring its response while varying the excitation frequency and the point of excitation [Stetson & Powell, 1965].

We assume that the hologram is recorded with the object stationary in its equilibrium position, and the system is adjusted so that the reconstructed image has the same intensity $I_0(x, y)$ as the object viewed through the hologram. The resultant intensity at any instant when the object is vibrating with an amplitude $\mathbf{L}(x, y)$ at a circular frequency ω is then

$$I(x, y, t) = I_0(x, y)\{1 - \cos[\mathbf{K} \cdot \mathbf{L}(x, y) \sin \omega t]\}. \tag{15.7}$$

Provided that the time-constant of the human eye (≈ 0.04 s) is much greater than the period of vibration, the observer sees the time-averaged intensity

$$\langle I(x, y) \rangle = I_0(x, y) \lim_{T \to \infty} \frac{1}{T} \int_0^T \{1 - \cos[\mathbf{K} \cdot \mathbf{L}(x, y) \sin \omega t]\} dt,$$
$$= I_0(x, y)\{1 - J_0[\mathbf{K} \cdot \mathbf{L}(x, y)]\}. \tag{15.8}$$

In this case, the characteristic function M_T is defined by the relation

$$|M_T|^2 = \{1 - J_0[\mathbf{K} \cdot \mathbf{L}]\}. \tag{15.9}$$

The function defined by (15.9) is plotted against $\Omega = \mathbf{K} \cdot \mathbf{L}$ in fig. 15.1(*b*). This corresponds to a dark field interferogram with only half the number of fringes seen with the time-average technique. However, in practice, because of varying phase shifts introduced by the photographic emulsion, it is difficult to get a dark field even with *in situ* processing. Typical fringe patterns obtained with the same object using the time-average and real-time techniques are shown in fig. 15.3 [Biedermann & Molin, 1970]. Most studies on vibrating objects involve a combination of these techniques. Usually, real-time holographic interferometry is used when searching for resonant modes, after which a time-average hologram can be

Fig. 15.2. Time-average interferograms showing some of the vibration modes of a model of an aircraft tail-fin. The corresponding vibration frequencies were (*a*) 670 Hz, (*b*) 894 Hz and (*c*) 3328 Hz [Abramson & Bjelkhagen, 1973].

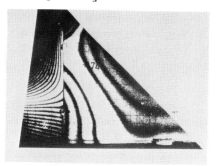

(*a*)

(*b*)

(*c*)

made for more accurate measurements of the vibration amplitude. Typical applications have been in studies of musical instruments [Ågren & Stetson, 1972], electromechanical devices such as loudspeakers [Chomat & Miler, 1973], turbine blades and aircraft structures [Bjelkhagen, 1974].

Very interesting possibilities have been opened up by the use of BSO (see section 7.7.2) for real-time phase conjugate imaging [Huignard, Herriau & Valentin, 1977]. A typical setup used to study the vibration modes of a loudspeaker is shown in fig. 15.4 [Marrakchi, Huignard & Herriau, 1980]. In this case, time-average fringes are formed because the recording medium integrates over the time constant τ involved in building up the hologram, which is long compared to the period of vibration. With such a setup it is possible to observe the changes in the time-average fringes in real time as the excitation frequency is changed.

15.1.3. Stroboscopic holographic interferometry

In stroboscopic holographic interferometry [Archbold & Ennos, 1968; Shajenko & Johnson, 1968; Watrasiewicz & Spicer, 1968] a hologram

Fig. 15.3. Fringe patterns obtained with the same vibrating object (a tin can) using (*a*) the time-average technique, and (*b*) real-time interference [Biedermann & Molin, 1970].

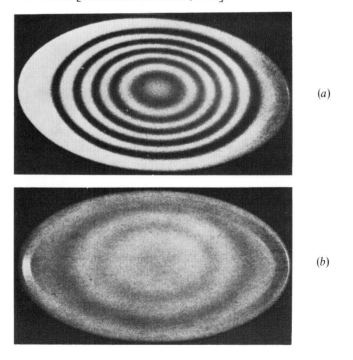

(*a*)

(*b*)

of a vibrating object is recorded using a sequence of light pulses that are triggered at times Δt_1 and Δt_2 during the vibration cycle. The characteristic function for an object displacement $\mathbf{L}(x, y)\sin \omega t$ is then

$$M_T = \exp[-i\mathbf{K}\cdot\mathbf{L}(x, y)\sin \omega \Delta t_1]$$
$$+\exp[-i\mathbf{K}\cdot\mathbf{L}(x, y)\sin \omega \Delta t_2], \tag{15.10}$$

so that the intensity in the image is

$$I(x, y) = I_0(x, y)\{1+\cos[\mathbf{K}\cdot\mathbf{L}(x, y)(\sin \omega \Delta t_1 - \sin \omega \Delta t_2)]\}. \tag{15.11}$$

The hologram is equivalent to a double-exposure hologram recorded while the object was in these two states of deformation, and the fringes have unit visibility irrespective of the vibrating amplitude. The phase of the vibration can be determined from a series of holograms made with different values of Δt_2, keeping Δt_1 fixed; alternatively, real-time observations can be made. More elaborate techniques employing pulsed illumination have been described by Vikram [1974a, b, 1975, 1976] for studying combinations of different types of motions.

15.1.4. Temporally modulated holography

Holography with a temporally modulated reference beam [Aleksoff, 1971] is a more sophisticated, but extremely powerful technique for the study of vibrations.

Fig. 15.4. Application of phase conjugation in BSO to the study of the modes of a vibrating loudspeaker [Marrakchi, Huignard & Herriau, 1980].

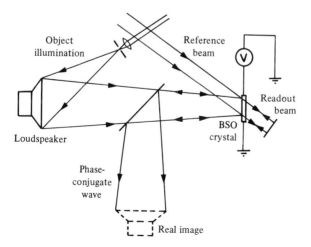

We assume, in this case, that the complex amplitude at any point on the hologram plane due to a point on the vibrating object is $f_o(t)o$, where $f_o(t) = \exp(-i\mathbf{K} \cdot \mathbf{L} \sin \omega t)$, while that due to the reference beam is $f_r(t)r$, where $f_r(t)$ describes the temporal modulation imposed on the reference beam. The intensity at the hologram plane at time t is then

$$I(t) = |f_o(t)o + f_r(t)r|^2. \tag{15.12}$$

As in section 15.1.1, we can consider this equivalent to recording a very large number of superposed holograms, one for each displaced position of the object. If we assume linear recording, as in (2.2), the amplitude transmittance of the hologram is

$$\mathbf{t} = \mathbf{t}_0 + \beta \int_0^T |f_o(t)o + f_r(t)r|^2 dt. \tag{15.13}$$

To view the reconstructed image, the hologram is illuminated by the unmodulated reference wave. The complex amplitude of the transmitted wave is then $u = r\mathbf{t}$, and the characteristic function can be written, apart from a constant factor, as

$$M_T = \frac{1}{T} \int_0^T f_r(t) f_o(t) dt. \tag{15.14}$$

The modulation of the reference beam can take different forms; some of the most interesting are discussed below, in sections 15.1.5–15.1.7.

15.1.5. Frequency translation

If the circular frequency of the reference wave is shifted by an amount $n\omega$, where n is an integer,

$$f_r(t) = \exp(-in\omega t), \tag{15.15}$$

and the characteristic function is

$$M_T = \lim_{T \to \infty} \frac{1}{T} \int_0^T \exp(-in\omega t) \exp(-i\mathbf{K} \cdot \mathbf{L} \sin \omega t) dt. \tag{15.16}$$

We then make use of the identity

$$\exp(iz \sin \phi) = \sum_{m=-\infty}^{\infty} J_m(z) \exp(im\phi), \tag{15.17}$$

and reverse the order of integration and summation to rewrite (15.16) as

$$M_T = \sum_{m=-\infty}^{\infty} J_m(\mathbf{K} \cdot \mathbf{L}) \lim_{T \to \infty} \frac{1}{T} \int_0^T \exp[-i(m-n)\omega t] dt. \tag{15.18}$$

However, when $T \gg 2\pi/\omega$,

$$\frac{1}{T} \int_0^T \exp[-i(m-n)\omega t] dt = \delta(m-n), \tag{15.19}$$

so that

$$M_T = J_n(\mathbf{K} \cdot \mathbf{L}),$$ (15.20)

and the intensity in the image is

$$I(x, y) = I_0(x, y)J_n^2[\mathbf{K} \cdot \mathbf{L}(x, y)].$$ (15.21)

Frequency translation can be used to obtain greatly increased sensitivity for the detection of small vibration amplitudes [Zambuto & Fischer, 1973]. As pointed out in section 15.1.1, with time-average fringes, the intensity is proportional to $J_0^2(\mathbf{K} \cdot \mathbf{L})$ and exhibits very little change for small amplitudes. On the other hand, with frequency translation, when $M_T = J_1(\mathbf{K} \cdot \mathbf{L})$, a dark field with fringes whose intensity is proportional to $(\mathbf{K} \cdot \mathbf{L})^2$ is obtained.

Frequency translation can also be used to obtain reduced sensitivity for measurements of large vibration amplitudes since, from (15.20), the number of fringes formed decreases as n is increased.

15.1.6. Amplitude modulation

In this case the amplitude of the reference wave is modulated at the same frequency as the vibrating object, so that $f_r(t) = \cos(\omega t - \psi)$, where ψ is the phase difference with reference to the vibrating object point $P(x, y)$. When the hologram is illuminated with an unmodulated reference wave, the intensity in the image is

$$I(x, y) = I_0(x, y)J_1^2[\mathbf{K} \cdot \mathbf{L}(x, y)]\cos^2 \psi.$$ (15.22)

A series of holograms recorded with different values of ψ can then be used to map the variations in phase over the vibrating object [Takai, Yamada & Idogawa, 1976].

15.1.7. Phase modulation

Phase information can also be obtained by sinusoidal phase modulation of the reference beam at the frequency of the vibrating object [Neumann, Jacobson & Brown, 1970]. In this case $f_r(t) = \exp(if \sin \omega t)$, where f is the amplitude of the cyclic phase shift. If the vibration at a point $P(x, y)$ on the object has a phase ψ_0, the characteristic function is

$$M_T = J_0\{[(\mathbf{K} \cdot \mathbf{L})^2 + f^2 - 2(\mathbf{K} \cdot \mathbf{L})f \cos \psi_0]^{1/2}\}.$$ (15.23)

Techniques for obtaining useful phase information using a minimum number of interferograms have been described by Levitt & Stetson [1976] and by Vikram [1977].

15.1.8. Holographic subtraction

This is a relatively simple technique [Hariharan, 1973] which involves making two equal exposures. During the first the object is at rest, while during the second it is made to vibrate after introducing a phase shift of π in one of the beams. The reconstructed amplitude due to the second exposure is subtracted from that due to the first, resulting in an interference pattern in which the intensity distribution is given by the relation

$$I(x, y) = I_0(x, y)\{1 - J_0[\mathbf{K} \cdot \mathbf{L}(x, y)]\}^2. \tag{15.24}$$

The characteristic function is the square of that for real-time fringes defined by (15.9) and shown in fig. 15.1(*b*). The number and position of the fringes is the same, but their visibility is higher.

With the double-exposure technique it is relatively easy to get a perfectly dark field, giving a useful increase in sensitivity for small vibration amplitudes. This method has been extended by using weighted subtraction to generate contour lines of equal vibration amplitude at any given small level [Sato, Ogawa & Ueda, 1974]. Other applications, including measurement of vibrations of large amplitude, separation of the effects of simultaneous uniform motion and the study of periodic nonsinusoidal vibrations, have been listed by Hariharan [1976*b*].

15.1.9. **Time-average holography of nonsinusoidal motions**

We consider, in the first instance, separable motions for which the displacements of all the points on the surface of the object can be written as the product of a spatially varying vector amplitude $\mathbf{L}(x, y)$ and a single function of time $f(t)$.

The characteristic function is then

$$M_T = \lim_{T \to \infty} \frac{1}{T} \int_0^T \exp[-i\mathbf{K} \cdot \mathbf{L}f(t)]dt, \tag{15.25}$$

where, as before, $\mathbf{K}$ is the sensitivity vector; this expression has been evaluated for various types of motion by Zambuto & Lurie [1970]; Janta & Miler [1972]; Stetson [1972*a*] and Gupta & Singh [1975*a, b*, 1976].

The analysis can be facilitated in some cases by an interpretation of the characteristic function suggested by Stetson [1971] according to which the reconstructed wavefront can be considered as the ensemble average of the object wavefronts recorded by the hologram. Accordingly (15.25) can be written as

$$M_T = \int_{-\infty}^{\infty} p(f) \exp(i\Omega f)df, \tag{15.26}$$

where $\Omega = \mathbf{K} \cdot \mathbf{L}$ and $p(f)$ is a probability density function.

More complex vibrations can be considered as the sum of several separable motions, so that the characteristic function can be written as

$$M_T = \lim_{T \to \infty} \frac{1}{T} \int_0^T \exp\left[-i \sum_{n=1}^{N} \Omega_n f_n(t) \right] dt, \tag{15.27}$$

where $\Omega_n = \mathbf{K} \cdot \mathbf{L}_n$.

In this case, if the motions are at irrationally related frequencies (temporally independent motions), the characteristic function is merely the product of the characteristic functions of the motions taken individually.

Vibrations at rationally related frequencies result in more complicated characteristic functions. Such vibrations have been studied by Molin & Stetson [1969], Stetson [1970c], Stetson & Taylor [1971], Wilson [1970, 1971], Wilson & Strope [1970], Dallas & Lohmann [1975] and Tonin & Bies [1978].

A useful approximation for the characteristic function can often be obtained in such cases by using the method of stationary phase [Stetson, 1972b]. This is equivalent to assuming that the time-average fringes are formed mainly by interference between the reconstructed wavefronts corresponding to those positions of the object at which it dwells for a relatively long time; this is, in fact, a simple physical explanation of the formation of time-average fringes.

15.1.10. Localization of the fringes with vibrating objects

Molin & Stetson [1970a, b] have shown experimentally that for separable motions the localization of the fringes follows the same pattern as with a double-exposure hologram for a displacement of the same nature. The situation is not so simple for nonseparable motions, except for independent motions, where the corresponding fringe systems localize independently [Molin & Stetson, 1971]. With mutually dependent vibrations, the fringes are not well localized and can usually be observed only with a properly oriented slit aperture. However, with temporally orthogonal vibrations, the fringes have high visibility where the zero-order fringe of one component intersects the plane of localization of the other.

15.2. Holographic photoelasticity

As is well known, information on the stresses in a model made of a material which becomes birefringent when it is stressed can be obtained by studying the state of polarization of the light transmitted by it. However, conventional photoelastic measurements only give the isochromatics, the loci of points corresponding to constant values of $(\sigma_1 - \sigma_2)$, the difference in the principal stresses.

Favre [1929] showed that interferometric measurements with unpolarized light can give a set of fringes, called isopachics, corresponding to variations in the thickness of the model. These fringes are the loci of points for which $(\sigma_1 + \sigma_2)$ the sum of the principal stresses is constant. Subsequently, Nisida & Saito [1964] developed an interferometric method which gives $(\sigma_1 - \sigma_2)$ and $(\sigma_1 + \sigma_2)$ simultaneously. However, both these methods require a test specimen of high optical quality.

Fourney [1968] and Hovanesian, Brcic & Powell [1968] were able to eliminate this requirement by applying holography to photoelastic measurements. Using a double-exposure technique, they obtained a combined isopachic–isochromatic fringe pattern which could be related to the stress distribution in the model [Fourney & Mate, 1970]. Unfortunately, the analysis of this pattern is not very easy, and ambiguities can arise [Holloway & Johnson, 1971; Sanford & Durelli, 1971]. Because of these difficulties, methods have been developed which make it possible to obtain separate isochromatic and isopachic patterns.

15.2.1. Isochromatics

A hologram of the stressed model recorded with coherent light which is either unpolarized or circularly polarized reconstructs the isochromatics.

This occurs because the light incident at a point P on the model, in the setup in fig. 15.5, can be resolved into two orthogonally polarized

Fig. 15.5. Holographic system used to obtain the isochromatics of a photoelastic model [O'Regan & Dudderar, 1971].

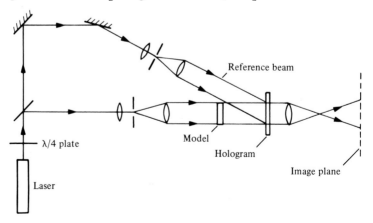

components, with electric vectors parallel to the principal stresses at this point, whose complex amplitudes can be written as $o_1 = \exp(-i\phi_1)$ and $o_2 = \exp(-i\phi_2)$, respectively. The reference wave can also be resolved into two similar orthogonal components r_1 and r_2. Interference takes place only between o_1 and r_1 and o_2 and r_2, respectively, so that two independent holograms are recorded on the same plate.

When the processed plate is illuminated once again by the reference beam, it reconstructs two waves whose complex amplitudes are proportional to o_1 and o_2, respectively. However, these are now able to interfere, exactly as in double-exposure holographic interferometry, since their states of polarization are the same. The intensity distribution in the image is then

$$I = |o_1 + o_2|^2,$$
$$= 2[1 + \cos(\phi_1 - \phi_2)]. \tag{15.28}$$

The phase difference between the two waves is

$$\phi_1 - \phi_2 = (2\pi/\lambda)(n_1 - n_2)(d + \Delta d), \tag{15.29}$$

where n_1 and n_2 are the principal refractive indices of the material when it is stressed, d is the thickness of the unstressed model and Δd is the change in its thickness.

In addition, n_1 and n_2 are related to n_0, the refractive index of the unstressed material, and σ_1 and σ_2, the principal stresses, by the Maxwell–Neumann equations

$$n_1 - n_0 = A\sigma_1 + B\sigma_2, \tag{15.30}$$
$$n_2 - n_0 = B\sigma_1 + A\sigma_2, \tag{15.31}$$

where A and B are the stress-optical coefficients of the material. Accordingly,

$$n_1 - n_2 = (A - B)(\sigma_1 - \sigma_2). \tag{15.32}$$

Since $\Delta d \ll d$, we then have from (15.28), (15.29) and (15.32),

$$I = 2\{1 + \cos[(2\pi/\lambda)(A - B)(\sigma_1 - \sigma_2)d]\}. \tag{15.33}$$

15.2.2. Isopachics

In order to obtain the isopachics, it is necessary to make two exposures on the hologram plate, one with a stress applied to the model and the other without any stress applied to it. The change in thickness of the model at any point is then linearly proportional to the sum of the principal stresses at this point and is given by the relation

$$\Delta d = -d(\nu/E)(\sigma_1 + \sigma_2), \tag{15.34}$$

where v and E are the Poisson's ratio and the Young's modulus, respectively. In this case, if the model did not exhibit any birefringence, the image reconstructed by the hologram would exhibit fringes corresponding to the isopachics. As it is, because of the birefringence of the model, a complicated fringe system is observed arising from the interaction of the two fringe systems.

The effects of the stress-induced birefringence can be eliminated if, as shown in fig. 15.6, the object beam is made to traverse the model once again after passing through an optical rotator (a Faraday cell) which rotates the axes of polarization by 90° [Chau, 1968; O'Regan & Dudderar, 1971; Chatelain, 1973]. As a result, the vibration which propagates along the fast axis on the outward journey returns along the slow axis and *vice versa*, with the result that the phase difference between them due to the stress-induced birefringence is cancelled.

In this case, the effects of the change in thickness of the model can be neglected since they are relatively small compared to the effects due to the changes in the refractive index. The optical path through the stressed model can then be taken as $(n_1 + n_2)d$, while that through the unstressed object is $2n_0 d$. Accordingly, the intensity in the image reconstructed by the double-

Fig. 15.6. Holographic system used to obtain the isopachics of a photoelastic model [O'Regan & Dudderar, 1971].

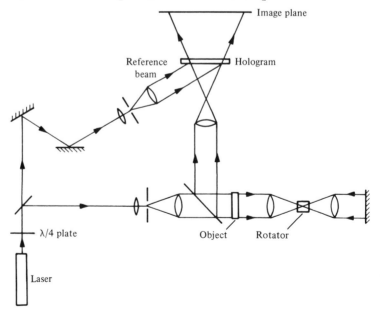

exposed hologram can be written as

$$I = 2I_0[1 + \cos(\phi - \phi_0)], \tag{15.35}$$

where $(\phi - \phi_0)$, the phase difference between the two reconstructed waves, is given by the relation

$$\phi - \phi_0 \approx (2\pi/\lambda)(n_1 + n_2 - 2n_0)d. \tag{15.36}$$

Now, from (15.30) and (15.31),

$$(n_1 + n_2 - 2n_0) = (A + B)(\sigma_1 + \sigma_2). \tag{15.37}$$

Accordingly the intensity of the reconstructed image is

$$I = 2\{1 + \cos[(2\pi/\lambda)(A + B)(\sigma_1 + \sigma_2)d]\}, \tag{15.38}$$

which corresponds to an isopachic fringe pattern.

Isochromatic and isopachic fringes obtained with a typical test object (an Araldite disc in diametral compression) using these techniques are shown in fig. 15.7 [Chatelain, 1973].

15.2.3. Combined displays

While the isochromatic and isopachic patterns can be obtained separately in this fashion, it is also possible to combine the two, making use of the multiplexing capabilities of holograms. An optical system using two reference beams has been described by Assa & Betser [1974] with which it is possible to obtain either of the patterns without interference from the other.

Another method of separating the isochromatic and isopachic fringes using a simple real-time holographic interferometer has been described by Hovanesian [1974]. A combined fringe pattern identical to that obtained by Nisida & Saito [1964] with an interferometer can also be obtained in a

Fig. 15.7. Holographic interference fringes obtained with a photoelastic model (an Araldite disc in diametral compression) showing (*a*) the isochromatics, (*b*) the isopachics [Chatelain, 1973].

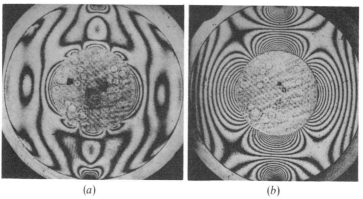

(*a*)   (*b*)

holographic system if a volume diffuser is used to depolarize the object beam [Kubo & Nagata, 1976*a, b*; Ebbeni, Coenen & Hermanne, 1976]. Sandwich holography can also be used for this purpose [Uozato & Nagata, 1977].

15.3.   Holographic contouring

Holographic interferometry can produce an image of a three-dimensional object modulated by a fringe pattern corresponding to contours of constant elevation with respect to a reference plane.

15.3.1.   Two-wavelength holographic contouring

In this method two holograms of the object are recorded using light of two different wavelengths [Haines & Hildebrand, 1965; Hildebrand & Haines, 1966, 1967].

To minimize the transverse displacement between the reconstructed images and obtain plane contouring surfaces, a telecentric system is used, as shown in fig. 15.8, to image the object on the hologram, and a plane wave is used to illuminate the object [Zelenka & Varner, 1968]. Another plane wave making an equal but opposite angle with the axis of the optical system is used as the reference wave. Two exposures are made with light of two different wavelengths, λ_1 and λ_2. After processing, the hologram is replaced in its original position and illuminated with one of the wavelengths, say λ_2.

Since the holograms are recorded and reconstructed with a collimated reference beam, (3.17) and (3.20) show that the two reconstructed images have the same magnification (unity). In addition, since they are formed close to the hologram plane, (3.38) shows that the lateral displacement is

Fig. 15.8. Optical system for two-wavelength holographic contouring [Zelenka & Varner, 1968].

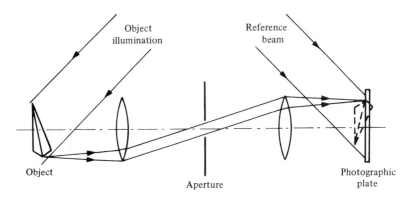

negligible. However, interference fringes are seen due to the axial displacement of one image with respect to the other, which from (3.39) is given by the relation

$$\Delta z = z(\lambda_1 - \lambda_2)/\lambda_1. \tag{15.39}$$

Since the telecentric system limits the rays forming the image to a small angle with the axis, successive fringes correspond to changes in Δz of $\lambda_2/2$, or, when λ_1 and λ_2 are not very different, to increments of z given by the relation

$$\delta z \approx \lambda^2/2\Delta\lambda. \tag{15.40}$$

A fairly wide range of contour intervals can be obtained with pairs of lines from an Ar^+ laser; thus, the two lines at $\lambda = 514$ nm and 488 nm give a contour interval δz of approximately 5 μm, while those at $\lambda = 488$ nm and 477 nm give a contour interval of approximately 10 μm.

With a dye laser, the contour interval can be varied continuously. It is also possible to make the wavelength difference very small. In the latter case the criteria for good fringe visibility can be satisfied more easily and simpler optical systems can be used [Friesem & Levy, 1976]. Photothermoplastics as well as BSO have been used as recording media for two-wavelength contouring [Leung, Lee, Bernal & Wyant, 1979; Küchel & Tiziani, 1981]; the latter permits real-time viewing of the contours.

15.3.2. Two-refractive-index contouring

The setup for this method [Tsuruta, Shiotake, Tsujiuchi & Matsuda, 1967; Zelenka & Varner, 1969], which requires only a single laser wavelength, is shown in fig. 15.9. The object is placed in a cell with a plane glass window and viewed through a telecentric system. A beam splitter is used to illuminate the object with a plane wave along the axis of the optical system. The hologram plate is located near the stop of the telecentric system.

Two holograms are recorded on the same plate, with the cell filled with fluids having refractive indices n_1 and n_2 respectively. Since the line of sight is normal to the window, there is no lateral displacement of the images; however, one of the images is longitudinally displaced with respect to the other by an amount

$$\Delta z = (n_1 - n_2)z, \tag{15.41}$$

where z is the distance from the window to the surface of the object.

Accordingly, when the hologram is replaced in the same position and illuminated once again by the same reference beam, successive fringes in the reconstructed image correspond to increments of z given by the relation

$$\delta z = \lambda/2|n_1 - n_2|. \qquad (15.42)$$

The contouring interval δz can be varied from about 1 μm to 300 μm by using air and a liquid or a combination of liquids, while combinations of gases can be used to obtain even larger contour intervals [Marrone & Ribbens, 1975].

In practice, two-refractive-index contouring is by far the simpler and more flexible of the two techniques. Its only disadvantage is the need to immerse the object in a suitable liquid when contours closer than about 300 μm are needed.

Fig. 15.9. Optical system for two-index holographic contouring [Zelenka & Varner, 1969].

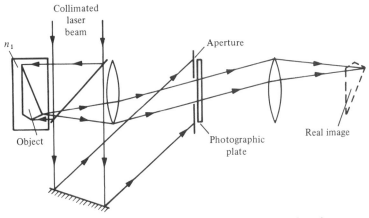

Fig. 15.10. Depth contours obtained by (*a*) the two-wavelength method (interval 9.25 μm); (*b*) the two-index method (interval 11.8 μm) [Zelenka & Varner, 1969].

(*a*)   (*b*)

Typical contours obtained with both these techniques using the same object are presented in fig. 15.10.

15.3.3. Contouring by changing the angle of illumination

A simple method of holographic contouring is to make a double-exposure hologram in which the point source illuminating the object is moved slightly sideways between the two exposures. In this case, the contouring surfaces consist of a set of hyperboloids of revolution with the two positions of the source as their common foci. Plane surfaces can be obtained if the dimensions of the object are small compared to the distance of the source, or if collimated illumination is used along with a telecentric imaging system.

Ordinarily, to obtain contouring surfaces normal to the line of sight with this technique, the beam illuminating the object must also be normal to the line of sight [Menzel, 1974]. However, contouring surfaces at any desired angle can be generated by translating the object between the two exposures in a properly chosen direction [Abramson, 1976b]. This can be combined with a displacement of the illuminating source [De Mattia & Fossati-Bellani, 1978]. A convenient alternative is the use of a sandwich hologram [Abramson, 1976a]; the contouring surfaces can then be made to assume any desired orientation by tilting the sandwich through a small angle.

15.3.4. Contouring with reflection holograms

A simple contouring technique using reflection holograms has been described by Henshaw & Ezekiel [1974]. This also has the advantage that the contoured image can be viewed with white light.

The optical arrangement is shown in fig. 15.11 and is similar to that used for 'piggyback' holograms (see section 5.1). The hologram plate is clamped

Fig. 15.11. Setup for contouring using reflection holography [Henshaw & Ezekiel, 1974].

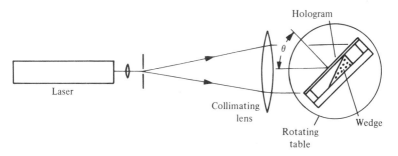

to the object and illuminated by a collimated beam. Two exposures are
made with the light incident on the hologram plate at angles θ_1 and θ_2
respectively. The contour spacing in the reconstructed image is then given
by the relation

$$\delta z = \lambda \cos \theta_0 / 2 \sin (\langle \theta \rangle - \theta_0) \sin (\Delta \theta / 2), \tag{15.43}$$

where θ_0 is the angle of viewing, $\langle \theta \rangle = (\theta_1 + \theta_2)/2$ and $\Delta \theta = \theta_1 - \theta_2$.

15.3.5. Photogrammetric contouring

In contouring techniques using holographic interference, the
contour fringes formed are localized on or near the surface of the virtual
image. If the object has appreciable depth, a picture of the fringes taken with
a camera gives a perspective projection of the fringes instead of a true
orthoscopic contour map. In such cases, there are advantages in using
techniques in which the contours are plotted by observations on the image
with an optical system having a very limited depth of focus [Stetson, 1968b;
Gara, Majkowski & Stapleton, 1973].

A more convenient way is to use a stereo plotter of the type developed for
photogrammetry [Balasubramanian, 1975].

In these, a self-illuminated dot (the tip of an optical fibre), attached to a

Fig. 15.12. Optical system used to produce a focused-image
holographic stereo model [Balasubramanian, 1975].

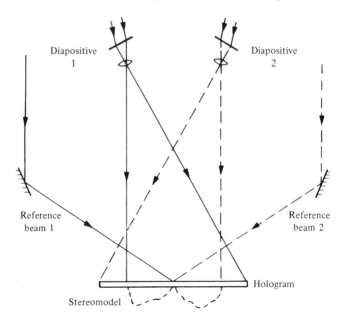

three-coordinate measuring device, is kept in apparent contact with the image while it is moved over a plane; it is then possible to plot contours (or profiles) directly. The multiplicity of perspectives available with a hologram makes this setting easier than with a stereo model and increases the precision. Settings can be made in the horizontal plane with an accuracy of 0.025 mm, and in depth with an accuracy of 0.75 mm.

If the object is too large to record a hologram of it, it is still possible to generate a holographic stereo model. This involves recording, on a single plate, holograms of two photographs of the object taken from the proper relative orientation. For this, the overlapped images can be projected on to a rear projection screen and holograms recorded of these projected images; however, a better method is to project the overlapped images directly on the hologram recording plate, as shown in fig. 15.12, to produce a focused-image holographic stereo model.

A1

The Fourier transform, convolution and correlation

A1.1.   The Fourier transform

The one-dimensional Fourier transform is widely used in the study of time-varying functions in the frequency domain [Papoulis, 1962; Bracewell, 1965]. Similarly, a function of two orthogonal spatial coordinates can be expressed by means of a two-dimensional Fourier transform, as a function of two orthogonal spatial frequencies [Goodman, 1968].

The two-dimensional Fourier transform of $g(x, y)$ is defined as

$$\mathscr{F}\{g(x, y)\} = \int_{-\infty}^{\infty} \int_{-\infty}^{\infty} g(x, y) \exp[-i2\pi(\xi x + \eta y)] \mathrm{d}x \mathrm{d}y,$$
$$= G(\xi, \eta). \tag{A1.1}$$

Similarly, the inverse Fourier transform of $G(\xi, \eta)$ is defined as

$$\mathscr{F}^{-1}\{G(\xi, \eta)\} = \int_{-\infty}^{\infty} \int_{-\infty}^{\infty} G(\xi, \eta) \exp[i2\pi(\xi x + \eta y)] \mathrm{d}\xi \mathrm{d}\eta,$$
$$= g(x, y). \tag{A1.2}$$

These relationships can be written symbolically as

$$g(x, y) \leftrightarrow G(\xi, \eta). \tag{A1.3}$$

The use of the Fourier transform permits studying the structure of an image in the spatial frequency domain and provides a useful insight into the working of an optical system. This is because the transform effectively decomposes a light wave into component plane waves whose direction cosines correspond to its spatial frequencies. The propagation of these components can then be analysed in a relatively simple fashion, as discussed in Appendix A2.

Some of the more important properties of the Fourier transform are summarized below.

a.   Linearity theorem

$$\mathscr{F}\{ag(x, y) + bh(x, y)\} = aG(\xi, \eta) + bH(\xi, \eta), \tag{A1.4}$$

where a and b are constants, and $g(x, y) \leftrightarrow G(\xi, \eta)$, $h(x, y) \leftrightarrow H(\xi, \eta)$.

b.   **Shift theorem**

$$\mathscr{F}\{g(x-a, y-b)\} = G(\xi, \eta)\exp[-i2\pi(\xi a + \eta b)]. \qquad (A1.5)$$

c.   **Similarity theorem**

$$\mathscr{F}\{g(ax, by)\} = (1/|ab|)G(\xi/a, \eta/b). \qquad (A1.6)$$

d.   **Rayleigh's theorem**

$$\int_{-\infty}^{\infty}\int_{-\infty}^{\infty} |g(x, y)|^2 dxdy = \int_{-\infty}^{\infty}\int_{-\infty}^{\infty} |G(\xi, \eta)|^2 d\xi d\eta. \qquad (A1.7)$$

A1.2.   Convolution and correlation

The convolution of two functions $g(x)$ and $h(x)$ is defined as

$$f(x) = \int_{-\infty}^{\infty} g(u)h(x-u)du. \qquad (A1.8)$$

This can be written as

$$f(x) = g(x) * h(x), \qquad (A1.9)$$

where the symbol $*$ denotes the convolution operation. The physical interpretation of the convolution operation is simple. The inverted shifted function $h(x-u)$ is multiplied by $g(u)$; the area under the curve $g(u)h(x-u)$ then gives the value of $f(x)$.

The convolution operation in two dimensions is defined as

$$f(x, y) = \int_{-\infty}^{\infty}\int_{-\infty}^{\infty} g(u, v)h(x-u, y-v)dudv, \qquad (A1.10)$$

which can also be written as

$$f(x, y) = g(x, y) * h(x, y). \qquad (A1.11)$$

A function frequently used in conjunction with the convolution operation is the Dirac delta function $\delta(x, y)$. By definition, convolution of a function with the delta function yields the original function, so that

$$\int_{-\infty}^{\infty}\int_{-\infty}^{\infty} f(u, v)\delta(x-u, y-v)dudv = f(x, y) \qquad (A1.12)$$

The delta function can be shown to take the values

$$\delta(x, y) = \infty \quad (x=0, \text{ and } y=0)$$
$$\delta(x, y) = 0 \quad (x \neq 0, \text{ or } y \neq 0) \qquad (A1.13)$$

and its integral is unity.

The cross-correlation of two functions $g(x, y)$ and $h(x, y)$, is

$$c(x, y) = \int_{-\infty}^{\infty}\int_{-\infty}^{\infty} g^*(u, v)h(x+u, y+v)dudv, \qquad (A1.14)$$

where $g^*(u, v)$ is the complex conjugate of $g(u, v)$. This can be written as

$$c(x, y) = g(x, y) \star h(x, y), \tag{A1.15}$$

where the symbol $\star$ denotes the correlation operation. A comparison with (A1.10) shows that the cross-correlation can also be expressed as a convolution

$$c(x, y) = g^*(x, y) * h(-x, -y). \tag{A1.16}$$

The autocorrelation of a function $g(x, y)$ is then

$$
\begin{aligned}
a(x, y) &= \int_{-\infty}^{\infty} \int_{-\infty}^{\infty} g^*(u, v)g(x+u, y+v)dudv, \\
&= g(x, y) \star g(x, y).
\end{aligned} \tag{A1.17}
$$

Some useful results which follow are listed below.

a. **The convolution theorem**

If $g(x, y) \leftrightarrow G(\xi, \eta)$ and $h(x, y) \leftrightarrow H(\xi, \eta)$,

$$\mathscr{F}\{g(x, y) * h(x, y)\} = G(\xi, \eta)H(\xi, \eta). \tag{A1.18}$$

b. **The autocorrelation (Wiener–Khinchin) theorem**

$$\mathscr{F}\{g(x, y) \star g(x, y)\} = |G(\xi, \eta)|^2. \tag{A1.19}$$

A1.3.   Random functions

The correlation function is very useful in the study of randomly varying quantities [Papoulis, 1965]. Since, in such a case, an integral such as (A1.14) would be infinite, the cross-correlation of two stationary random functions, typically functions of time, $g(t)$ and $h(t)$, is written as

$$
\begin{aligned}
R_{gh}(\tau) &= \lim_{T \to \infty} \frac{1}{2T} \int_{-T}^{T} g^*(t)h(t+\tau)dt, \\
&= \langle g^*(t)h(t+\tau) \rangle.
\end{aligned} \tag{A1.20}
$$

The autocorrelation function of one of them, say $g(t)$, is

$$R_{gg}(\tau) = \langle g^*(t)g(t+\tau) \rangle. \tag{A1.21}$$

If $g(t)$ is, for example, the time-varying electric field at a point due to an electromagnetic wave, the average power at this point is $R_{gg}(0)$.

The power spectrum $S(\omega)$ of a random function $g(t)$ is the Fourier transform of its autocorrelation

$$R_{gg}(\tau) \leftrightarrow S(\omega). \tag{A1.22}$$

Hence, if $R_{gg}(\tau)$ is sharply peaked, (in the limit, a delta function), $S(\omega)$ must extend to very high frequencies (and, in the limit, is a constant).

A1.4. Sampling and the discrete Fourier transform

The production of computer-generated holograms involves computation of the Fourier transform $G_s(\xi, \eta)$ of a sampled function $g_s(x, y)$ which is obtained by sampling the object wave $g(x, y)$ at intervals $(\Delta x, \Delta y)$. If, for simplicity, we consider the one-dimensional case, as shown in fig. A1.1, we can write

$$g_s(x) = g(x) \sum_{m=-\infty}^{\infty} \delta(x - m\Delta x), \qquad (A1.23)$$

and

$$G_s(\xi) = (1/\Delta x) \sum_{m=-\infty}^{\infty} G[\xi - (m/\Delta x)], \qquad (A1.24)$$

where $G(\xi) \leftrightarrow g(x)$.

As shown in fig. A1.2, the Fourier transform of the sampled function $g_s(x)$ is a regular series of repetitions of the Fourier transform of the original function $g(x)$, shifted in frequency space by successive intervals $\Delta \xi = (1/\Delta x)$. Overlap of the shifted Fourier transforms would normally result in aliasing

Fig. A1.1. The function $g_s(x)$ consists of an array of delta functions obtained by sampling the function $g(x)$ at intervals of Δx.

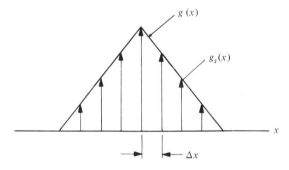

Fig. A1.2. The Fourier transform $G_s(\xi)$ of the sampled function $g_s(x)$ consists of a regular series of repetitions of $G(\xi)$, the Fourier transform of $g(x)$, shifted in frequency space by successive intervals $\Delta \xi = 1/\Delta x$.

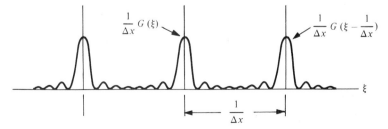

and make it impossible to recover $g(x)$ exactly. However, if the sampling interval for the object wave is chosen so that

$$\Delta x \leqslant 1/\xi_{max}, \tag{A1.25}$$

it is possible to recover $G(\xi)$ from $G_s(\xi)$. In the same manner, it is possible to recover $g(x)$ from the image reconstructed by a computed hologram, provided the sampling interval for it also is properly chosen.

For this, the values of the Fourier transform $G_s(\xi)$ are required at a series of points in the hologram plane taken at intervals of ξ given by the relation

$$\Delta \xi \leqslant 1/x_{max}, \tag{A1.26}$$

which is analogous to (A1.25). The values of $G_s(\xi)$ at these points can be obtained from the discrete Fourier transform

$$G_s(n/x_{max}) = (1/\xi_{max}) \sum_{m=-M/2}^{M/2} g(m/\xi_{max})$$
$$\times \exp(i2\pi mn/\xi_{max}x_{max}). \tag{A1.27}$$

A2

Wave propagation and diffraction

A2.1. Computation of phase across a spherical wavefront

Consider a point $S(x_1, y_1, 0)$ in the coordinate system shown in fig. A2.1 emitting light waves of wavelength λ. Their phase at a point $P(x, y, z)$ relative to that at the point $P_0(0, 0, z)$ can be calculated from the optical path difference and is

$$
\begin{aligned}
\phi &= (2\pi/\lambda)(SP - SP_0), \\
&= (2\pi/\lambda)\{[(x - x_1)^2 + (y - y_1)^2 + z^2]^{1/2} \\
&\quad - [x_1^2 + y_1^2 + z^2]^{1/2}\},
\end{aligned}
\tag{A2.1}
$$

$$
= (2\pi/\lambda)z\left\{\left[1 + \frac{(x - x_1)^2 + (y - y_1)^2}{z^2}\right]^{1/2} - \left[1 + \frac{x_1^2 + y_1^2}{z^2}\right]^{1/2}\right\}.
\tag{A2.2}
$$

If z is large compared to x_1, y_1, x and y, this can be written approximately, to the first order in z, as

$$
\phi = (2\pi/\lambda)(1/2z)(x^2 + y^2 - 2xx_1 - 2yy_1).
\tag{A2.3}
$$

Fig. A2.1. Coordinate system used to evaluate the Fresnel-Kirchhoff integral.

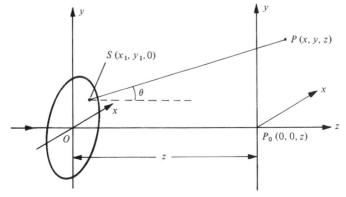

A2.2. The Fresnel–Kirchhoff integral

The Fresnel–Kirchhoff integral [O'Neill, 1963; Born & Wolf, 1980] states that if, as shown in fig. A2.1, a plane wave with an amplitude a is incident normally on an object located in the plane $z = 0$, with an amplitude transmittance $t(x_1, y_1)$, the complex amplitude at a point $P(x, y, z)$ is

$$
\begin{aligned}
a(x, y, z) = (\mathrm{i}a/\lambda) \int_{-\infty}^{\infty} \int_{-\infty}^{\infty} & t(x_1, y_1) \\
& \times \frac{\exp\{(-\mathrm{i}2\pi/\lambda)[(x-x_1)^2 + (y-y_1)^2 + z^2]^{1/2}\}}{[(x-x_1)^2 + (y-y_1)^2 + z^2]^{1/2}} \\
& \times \cos\theta \, \mathrm{d}x_1 \mathrm{d}y_1.
\end{aligned}
\tag{A2.4}
$$

If the point P is at a distance z much larger than $(x - x_1)$ and $(y - y_1)$, $\cos\theta \approx 1$, and (A2.4) can be written as

$$
\begin{aligned}
a(x, y, z) = (\mathrm{i}a/\lambda z) \int_{-\infty}^{\infty} \int_{-\infty}^{\infty} & t(x_1, y_1) \\
& \times \exp\{(-\mathrm{i}\pi/\lambda z)[(x-x_1)^2 + (y-y_1)^2]\} \mathrm{d}x_1 \mathrm{d}y_1,
\end{aligned}
\tag{A2.5}
$$

if we omit a factor $\exp(-\mathrm{i}2\pi/\lambda z)$, which only affects the overall phase. This can be expanded to give

$$
\begin{aligned}
a(x, y, z) = (\mathrm{i}a/\lambda z) \int_{-\infty}^{\infty} \int_{-\infty}^{\infty} & t(x_1, y_1) \exp[(-\mathrm{i}\pi/\lambda z)(x^2 + y^2)] \\
& \times \exp[(-\mathrm{i}\pi/\lambda z)(x_1^2 + y_1^2)] \\
& \times \exp\{\mathrm{i}2\pi(x_1(x/\lambda z) + y_1(y/\lambda z))\} \mathrm{d}x_1 \mathrm{d}y_1.
\end{aligned}
\tag{A2.6}
$$

Since the first exponential factor in the above expression is independent of x_1 and y_1, it can be taken outside the integral sign. In addition, if the distance to the plane of observation is large compared to the dimensions of the object, so that

$$
z \gg (x_1^2 + y_1^2)/\lambda,
\tag{A2.7}
$$

(the far-field condition), the second exponential factor is approximately equal to unity. If then, we set

$$
\begin{aligned}
\xi &= x/\lambda z, \\
\eta &= y/\lambda z,
\end{aligned}
\tag{A2.8}
$$

(A2.6) becomes

$$
\begin{aligned}
a(x, y, z) &= (\mathrm{i}a/\lambda z) \exp[(-\mathrm{i}\pi/\lambda z)(x^2 + y^2)] \\
& \quad \times \int_{-\infty}^{\infty} \int_{-\infty}^{\infty} t(x_1, y_1) \exp[\mathrm{i}2\pi(\xi x_1 + \eta y_1)] \mathrm{d}x_1 \mathrm{d}y_1, \\
&= (\mathrm{i}a/\lambda z) \exp[(-\mathrm{i}\pi/\lambda z)(x^2 + y^2)] T(\xi, \eta),
\end{aligned}
\tag{A2.9}
$$

where

$$\mathbf{t}(x_1, y_1) \leftrightarrow \mathbf{T}(\xi, \eta). \tag{A2.10}$$

It follows that the complex amplitude in the plane of observation is given by the Fourier transform of the amplitude transmittance of the object, multiplied by a spherical phase factor.

A2.3. The spherical lens

A thin convex lens illuminated by a collimated beam brings this beam to a focus at a distance f from the lens equal to its focal length. Assuming no absorption, the effect of the lens is merely to introduce a phase delay $\Delta\phi(x, y)$ which varies over the pupil and converts the wavefront from a plane wavefront to a spherical wavefront with its centre at the principal focus.

From (A2.3) this phase delay can be written as

$$\Delta\phi(x, y) = (\pi/\lambda f)(x^2 + y^2), \tag{A2.11}$$

so that a thin lens can be considered equivalent to a transparency with a complex amplitude transmittance

$$g(x, y) = \exp[(i\pi/\lambda f)(x^2 + y^2)]. \tag{A2.12}$$

If, now, a transparency with an amplitude transmittance $\mathbf{t}(x, y)$ is placed in front of the lens, the net transmitted amplitude is $\mathbf{t}(x, y)\exp[(i\pi/\lambda f)(x^2 + y^2)]$. The complex amplitude at a point (x_f, y_f) in the back focal plane of the lens can then be calculated from (A2.6) and is given by the relation

$$a_f(x_f, y_f) = (i/\lambda f)\exp[(-i\pi/\lambda f)(x_f^2 + y_f^2)]$$
$$\times \int_{-\infty}^{\infty} \int_{-\infty}^{\infty} \mathbf{t}(x, y)\exp\{i2\pi[x(x_f/\lambda f) + y(y_f/\lambda f)]\}dxdy.$$
$$\tag{A2.13}$$

A2.4. Fourier transformation by a lens

Consider the optical system shown in fig. A2.2 in which a plane wave with an amplitude a is incident normally on an object with an amplitude transmittance $\mathbf{t}(x_1, y_1)$ located in the front focal plane of a lens. From (A2.5) the complex amplitude a_l in the pupil of the lens is

$$a_l(x, y) = (ia/\lambda f) \int_{-\infty}^{\infty} \int_{-\infty}^{\infty} \mathbf{t}(x_1, y_1)$$
$$\times \exp\{(-i\pi/\lambda f)[(x - x_1)^2 + (y - y_1)^2]\}dx_1dy_1. \tag{A2.14}$$

From (A2.13), the complex amplitude in the back focal plane of the lens

Fig. A2.2. Fourier transformation by a lens.

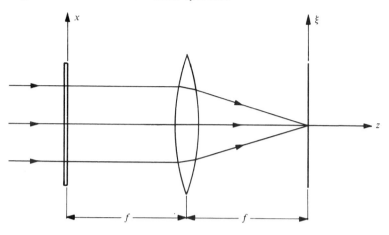

can then be written as

$$a_f(x_f, y_f) = (1/i\lambda f)\exp[(-i\pi/\lambda f)(x_f^2 + y_f^2)]$$
$$\times \int_{-\infty}^{\infty}\int_{-\infty}^{\infty} a_l(x, y)\exp\{i2\pi[x(x_f/\lambda f) + y(y_f/\lambda f)]\}\mathrm{d}x\mathrm{d}y,$$

$$(A2.15)$$

or as

$$a_f(\xi, \eta) = (1/i\lambda f)\exp[i\pi\lambda f(\xi^2 + \eta^2)]A_l(\xi, \eta), \qquad (A2.16)$$

where $\quad \xi = x_f/\lambda f, \qquad \eta = y_f/\lambda_f$

and $\quad a_l(x, y) \leftrightarrow A_l(\xi, \eta).$ $\qquad\qquad\qquad\qquad\qquad\qquad (A2.17)$

Now, (A2.14) can also be regarded as a convolution

$$a_l(x, y) = (ia/\lambda f)\mathbf{t}(x_1, y_1) * \exp[(-i\pi/\lambda f)(x_1^2 + y_1^2)], \qquad (A2.18)$$

so that, if we take the Fourier transforms of both sides,

$$A_l(\xi, \eta) = (ia/\lambda f)\mathbf{T}(\xi, \eta)\exp[-i\pi\lambda f(\xi^2 + \eta^2)]. \qquad (A2.19)$$

When (A2.19) is substituted in (A2.16), we obtain

$$a_f(\xi, \eta) = (a/\lambda^2 f^2)\mathbf{T}(\xi, \eta), \qquad\qquad\qquad (A2.20)$$

A3

Interference and coherence

A3.1. Interference

The time-varying electric field at any point due to a linearly polarized light wave can be represented by the analytic signal [Born & Wolf, 1980]

$$V(t) = \int_0^\infty a(v) \exp[-i\phi(v)] \exp(i2\pi vt), \tag{A3.1}$$

where $a(v)$ is the amplitude and $\phi(v)$ is the phase of a component with frequency v. For light waves of a single frequency, (A3.1) reduces to

$$V(t) = |a| \exp(-i\phi) \exp(i2\pi vt). \tag{A3.2}$$

In (A3.2) the factor $a = |a| \exp(-i\phi)$, which does not vary with time, is called the complex amplitude.

The optical intensity I at a point, which is defined as the time average of the amount of energy which crosses, in unit time, a unit area perpendicular to the energy flow, is obtained by multiplying the complex amplitude at this point by its complex conjugate, so that

$$I = |a|^2 = aa^*. \tag{A3.3}$$

The complex amplitude at any point due to a number of waves of the same frequency is obtained by summing the complex amplitudes of the individual waves. Thus,

$$a = a_1 + a_2 + a_3 + \ldots \tag{A3.4}$$

Accordingly, the intensity at any point due to the interference of two waves is

$$\begin{aligned} I &= |a_1 + a_2|^2, \\ &= |a_1|^2 + |a_2|^2 + a_1 a_2^* + a_1^* a_2, \\ &= I_1 + I_2 + 2(I_1 I_2)^{1/2} \cos(\phi_1 - \phi_2). \end{aligned} \tag{A3.5}$$

The visibility of the interference fringes is

$$\begin{aligned} \mathcal{V} &= (I_{max} - I_{min})/(I_{max} + I_{min}), \\ &= 2(I_1 I_2)^{1/2}/(I_1 + I_2). \end{aligned} \tag{A.36}$$

It should be noted that (A3.4) and (A3.5) involve the assumption that the

261

waves are polarized with their electric vectors parallel. If the two electric vectors make an angle ψ with each other, the resultant intensity in the interference pattern becomes

$$I = I_1 + I_2 + 2(I_1 I_2)^{1/2} \cos \psi \cos (\phi_1 - \phi_2). \tag{A3.7}$$

Obviously, the visibility of the fringes decreases as ψ increases, and drops to zero when $\psi = \pi/2$.

A3.2. Coherence

Another assumption implicit in (A3.4) and (A3.5) is that the light waves are perfectly coherent; this would be the case if they were all derived from a single point source emitting an infinitely long, continuous, monochromatic wave train. However, all real wave-fields only have a finite degree of coherence.

For the source S shown in fig. A3.1, if $V_1(t)$ and $V_2(t)$ are the analytic signals corresponding to the electric fields at P_1 and P_2, the complex degree of coherence $\gamma_{12}(\tau)$ for a time delay τ is defined as the normalized correlation of $V_1(t)$ and $V_2(t)$, which from (A1.20) and (A1.21) can be written as

$$\gamma_{12}(\tau) = \frac{\langle V_1(t+\tau)V_2^*(t)\rangle}{[\langle V_1(t)V_1^*(t)\rangle\langle V_2(t)V_2^*(t)\rangle]^{1/2}}. \tag{A3.8}$$

The physical significance of (A3.8) can be understood if the light waves are allowed to emerge through pinholes at P_1 and P_2 so that they form an interference pattern on a screen.

P_1 and P_2 can now be considered as two secondary sources, so that, from (A3.5), the intensity at Q is

$$\begin{aligned} I &= I_1 + I_2 + \langle V_1(t+\tau)V_2^*(t) + V_1^*(t+\tau)V_2(t)\rangle, \\ &= I_1 + I_2 + 2\text{Re}[\langle V_1(t+\tau)V_2^*(t)\rangle], \end{aligned} \tag{A3.9}$$

Fig. A3.1. Evaluation of the degree of coherence.

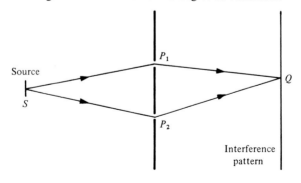

where I_1 and I_2 are the intensities at Q due to the sources P_1 and P_2 acting separately, and τ is the difference in transit time between the paths P_1Q and P_2Q.

From (A3.8), it follows that (A3.9) can be rewritten as

$$
\begin{aligned}
I &= I_1 + I_2 + 2(I_1 I_2)^{1/2} \mathrm{Re}[\gamma_{12}(\tau)], \\
&= I_1 + I_2 + 2(I_1 I_2)^{1/2} |\gamma_{12}(\tau)| \cos \phi_{12}(\tau),
\end{aligned}
\tag{A3.10}
$$

where $\phi_{12}(\tau)$ is the phase of $\gamma_{12}(\tau)$.

Interference fringes are produced by the variations in $\cos \phi_{12}(\tau)$ across the screen. If $I_1 = I_2$, the visibility of these fringes is then, from (A3.6)

$$
\mathcal{V} = |\gamma_{12}(\tau)|.
\tag{A3.11}
$$

The coherence of the field due to any actual light source can be studied under two headings.

A3.3. Spatial coherence

When the difference in the optical paths to P_1 and P_2 is small ($\tau \approx 0$), we are dealing essentially with the spatial coherence of the field. For a quasimonochromatic source consisting of a number of uncorrelated oscillators (a pinhole in front of a thermal source) illuminating two points in a plane at a relatively large distance from this source, the complex degree of coherence between the electric fields at these two points is given by the expression

$$
\gamma(x) = \frac{\langle V_1(t) V_2^*(t) \rangle}{[\langle V_1(t) V_1^*(t) \rangle \langle V_2(t) V_2^*(t) \rangle]^{1/2}}.
\tag{A3.12}
$$

The degree of coherence of the source is then given by the van Cittert–Zernike theorem as the normalized Fourier transform of the intensity distribution over the source.

A3.4. Temporal coherence

If the source is of very small dimensions (effectively a point source) but radiates over a finite range of wavelengths, we are dealing with the temporal coherence of the field. In this case, the complex degree of coherence depends only on τ, the difference in the transit times from the source to P_1 and P_2, and is given by the expression

$$
\gamma(\tau) = \frac{\langle V(t) V^*(t + \tau) \rangle}{\langle V(t) V^*(t) \rangle}.
\tag{A3.13}
$$

This can be transformed and rewritten as

$$
\gamma(\tau) = \mathcal{F}\{S(v)\} \left/ \int_{-\infty}^{\infty} S(v) \, dv, \right.
\tag{A3.13}
$$

where $S(v)$ is the frequency spectrum of the radiation.

From (A3.10) and (A3.11) the degree of temporal coherence is given by the visibility of the interference fringes as the optical path difference is varied. This leads to the concepts of coherence time and coherence length.

For a source of mean frequency v_0 with a bandwidth Δv, it can be shown that the visibility of the fringes drops to zero for a time difference $\Delta\tau$ given by the uncertainty relation

$$\Delta\tau\Delta v \approx 1. \tag{A3.15}$$

This time $\Delta\tau$ is called the coherence time of the radiation; the coherence length is defined as

$$\Delta l \approx c\Delta\tau,$$
$$\approx c/\Delta v = \lambda_0^2/\Delta\lambda,$$

where c is the speed of light, λ_0 is the mean wavelength and $\Delta\lambda$ is the range of wavelengths emitted by the source.

For interference fringes of good visibility to be obtained, the path difference must be small compared to the coherence length.

A4

Speckle

The problem of speckle arises as soon as any diffusely scattering object is illuminated by a highly coherent source, such as a laser, because most surfaces are extremely rough on a scale of light wavelengths. Each of the microscopic elements making up the surface gives rise to a coherent diffracted wave. However, the optical paths to neighbouring elements exhibit random differences which may amount to several wavelengths. Consequently, when the diffracted waves from these elements interfere with each other, a stationary granular pattern results, called a speckle pattern (see fig. A4.1). The general appearance of such speckle patterns is almost independent of the character of the surface, but the scale of the granularity increases with the viewing distance and the f-number of the viewing system.

The statistics of such speckle patterns have been analysed in detail by Goodman [1975]; some of the most important results are summarized in the next few sections.

Fig. A4.1. Speckle pattern observed when a diffusing surface is illuminated with coherent light.

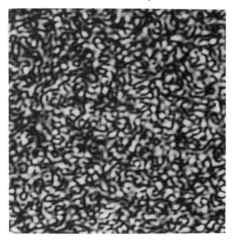

A4.1. First-order statistics of speckle patterns

The complex amplitude at any point in the far field of a diffuser illuminated by a coherent source as shown in fig. A4.2 is the sum of the complex amplitudes of the diffracted waves from all the individual elements on the object; it is given by the relation

$$a\exp(-i\phi)=\sum_{n=1}^{n} a_n\exp(-i\phi_n).$$ (A4.1)

If it is assumed that the moduli of all the individual complex amplitudes are equal, while the phase shifts are large enough that the remainders, after subtracting integral multiples of 2π, are uniformly distributed over the range from 0 to 2π, this reduces to the well-known random-walk problem. The joint probability density function of the real and imaginary parts of the complex amplitude is

$$p_{r,i}(a_{r,i})=(1/2\pi\sigma^2)\exp\left[-(a_r^2+a_i^2)/2\sigma^2\right],$$ (A4.2)

where σ^2 is a constant. The most common value of the modulus is zero, while the phase has a uniform circular distribution. It can then be shown that the probability density function of the intensity is the negative exponential distribution

$$p(I)=(1/2\sigma^2)\exp(-I/2\sigma^2).$$ (A4.3)

From (A4.3) it follows that the mean value of the intensity is

$$\langle I\rangle=2\sigma^2,$$ (A4.4)

while its second moment is

$$\langle I^2\rangle=2\langle I\rangle^2.$$ (A4.5)

Fig. A4.2. Coordinate system used to study the statistics of speckle patterns.

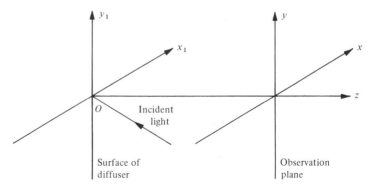

Accordingly, the variance of the intensity is

$$\sigma_I^2 = \langle I^2 \rangle - \langle I \rangle^2,$$
$$= \langle I \rangle^2. \tag{A4.6}$$

If the contrast of the speckle pattern is defined as

$$c = \sigma_I / \langle I \rangle, \tag{A4.7}$$

it is apparent that the contrast of a speckle pattern formed by perfectly coherent light is unity.

A4.2. Second-order statistics of speckle patterns

The two quantities of interest are the autocorrelation function and the power spectral density of the intensity distribution.

To evaluate these, we make use of the fact that the complex amplitude $A(x_1, y_1)$ at the scattering surface and the complex amplitude $a(x, y)$ at the observation plane are related by the Fresnel–Kirchhoff integral. The autocorrelation function of the intensity in the speckle pattern is then given by the relation

$$R_I(\Delta x, \Delta y) =$$

$$\langle I \rangle^2 \left\{ 1 + \left| \frac{\displaystyle\iint_{-\infty}^{\infty} |A(x_1, y_1)|^2 \exp\left[(i2\pi/\lambda z)(x_1\Delta x + y_1 \Delta y)\right] dx_1 dy_1}{\displaystyle\iint_{-\infty}^{\infty} |A(x_1, y_1)|^2 dx_1 dy_1} \right|^2 \right\} \tag{A4.8}$$

For a square scattering surface whose edges have a length L, the average dimensions of a speckle calculated from (A4.8) are

$$\delta x = \delta y = \lambda z / L. \tag{A4.9}$$

The power spectral density of the intensity distribution is given by the Fourier transform of the autocorrelation function $R_I(\Delta x, \Delta y)$ which is

$$S_I(s_x, s_y) = \langle I \rangle^2 \left\{ \delta(s_x, s_y) \right.$$

$$\left. + \frac{\displaystyle\iint_{-\infty}^{\infty} |A(x_1, y_1)|^2 |A(x_1 - \lambda z s_x, y_1 - \lambda z s_y)|^2 dx_1 dy_1}{\left[\displaystyle\iint_{-\infty}^{\infty} |A(x_1, y_1)|^2 dx_1 dy_1\right]^2} \right\} \tag{A4.10}$$

where s_x, s_y are spatial frequencies along the x and y axes, respectively. Apart from a delta function at zero spatial frequency which contains half the total power, this is the normalized autocorrelation function of the intensity distribution over the scattering surface.

A4.3.   Image speckle

As distinct from the direct scattered field, the statistics of the intensity fluctuations in the image of the scattering surface formed by a lens (or by a hologram) are dependent on the size of the imaging aperture, which acts as a low-pass filter. The entrance pupil of the optical system can be considered as being illuminated by the primary speckle pattern. This random field then appears in the exit pupil, so that the intensity fluctuations in the image can be obtained by treating the exit pupil as a rough object. In this case, for a circular pupil of radius ρ, the average size of the speckles in the image is, from (A4.8),

$$\delta x = \delta y = 0.61 \ \lambda f / \rho, \tag{A4.11}$$

where f is the focal length of the optical system.

A4.4.   Addition of speckle patterns

Two limiting cases are possible when two or more speckle patterns are superposed. The first is where the light fields are coherent and the amplitudes add. It can be shown that in this case the first-order statistics of the pattern remain unchanged.

The other case is where the light fields are incoherent and the intensities add. If two speckle patterns with equal average intensity $\langle I/2 \rangle$ are superposed in this manner, it can be shown that the probability density function of the intensity in the resulting pattern is

Fig. A4.3. Probability density functions of the irradiance in (a) a single speckle pattern with $\langle I \rangle = 1$, and (b) the incoherent sum of two speckle patterns with $\langle I_1 \rangle = \langle I_2 \rangle = 0.5$.

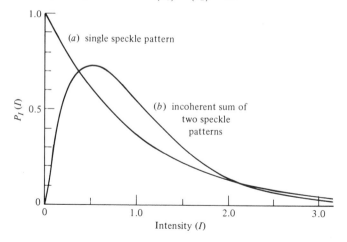

$$p(I) = (4I/\langle I \rangle^2) \exp(-2I/\langle I \rangle). \tag{A4.12}$$

As can be seen from fig. A4.3, this differs from the probability density function for a single speckle pattern with the same average intensity, which is also plotted for comparison, mainly in the elimination of most of the dark areas. As a result its contrast is only $2^{-1/2}$. If N speckle patterns with the same average intensity are superposed, the contrast of the resultant speckle pattern drops to $N^{-1/2}$.

A5

The H & D curve

The response of photographic materials to exposure to light is normally represented by a curve (known as the Hurter and Driffield, or H & D curve) in which the optical density of the material, after it has been developed and fixed, is plotted against the logarithm of the exposure given to it. The optical density D is defined by the relation

$$D = \log 1/\mathcal{T}, \qquad (A5.1)$$

where $\mathcal{T}$ is the transmittance of the material for intensity, while the exposure is defined as the product of I, the intensity of the light to which the material has been exposed, and T, the exposure time.

Typical H & D curves for two photographic materials are shown in fig. A5.1. As can be seen, the upper portion of the curves is a straight line. One of

Fig. A5.1. Typical H & D curves for Holotest 10E75 and 8E75 plates ($\lambda = 633$ nm).

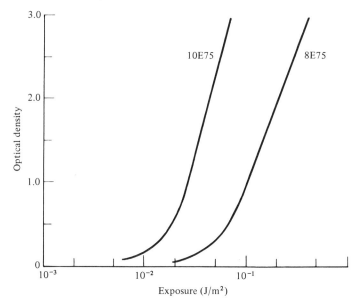

270

the parameters commonly used to characterize a photographic material is the value of γ, the slope of this section. It can be shown that, over this region,

$$\mathscr{T} \propto I^{-\gamma}. \tag{A5.2}$$

Although the H & D curve is used almost universally for photography, it is not a convenient way to specify the response of materials for holography, for which a curve showing $\mathbf{t}$, the amplitude transmittance of the material, as a function of the exposure is preferable. Since $\mathbf{t} = \mathscr{T}^{1/2}$, (A5.2) can be rewritten as

$$\mathbf{t} \propto I^{-\gamma/2}, \tag{A5.3}$$

over the linear portion of the H & D curve.

If a positive transparency is made from a photographic negative, we have, overall,

$$\gamma = -\gamma_n \gamma_p, \tag{A5.4}$$

where γ_n and γ_p are the slopes of the H & D curves for the negative and positive materials, respectively. In this case, for the final transparency,

$$\mathbf{t} \propto I^{\gamma_n \gamma_p/2}. \tag{A5.5}$$

It should be noted that the value of γ only determines the response of a photographic material on a macroscopic scale; its response on a microscopic scale also involves the modulation transfer function (see section 6.2).

Bibliography

Abramson, N. (1981). *The Making and Evaluation of Holograms*. London: Academic Press.

Butters, J. N. (1971). *Holography and its Technology*. London: Peter Peregrinus Ltd.

Caulfield, H. J. & Lu. S. (1970). *The Applications of Holography*. New York: Wiley-Interscience.

Caulfield, H. J. ed. (1979). *Handbook of Optical Holography*. New York: Academic Press.

Collier, R. J., Burckhardt, C. B. & Lin, L. H. (1971). *Optical Holography*. New York: Academic Press.

De Velis, J. B. & Reynolds, G. O. (1967). *Theory and Applications of Holography*. Reading: Addison-Wesley Publishing Co.

Erf, R. K. ed. (1974). *Holographic Non-destructive Testing*. New York: Academic Press.

Françon, M. (1969). *Holographie*. Paris: Masson et Cie.

Lehmann, M. (1970). *Holography: Theory and Practice*. London: Focal Press.

Menzel, E., Mirandé, W. & Weingärtner, I. (1973). *Fourier-Optik und Holographie*. Wien: Springer-Verlag.

Okoshi, T. (1976). *Three-dimensional Imaging Techniques*. New York: Academic Press.

Ostrovsky, Yu. I. (1977). *Holography and its Applications*. (Transl. G. Leib). Moscow: Mir Publishers.

Ostrovsky, Yu. I., Butusov, M. M. & Ostrovskaya, G. V. (1980). *Interferometry by Holography*. Berlin: Springer-Verlag.

Saxby, G. (1980). *Holograms*. London: Focal Press.

Schumann, W. & Dubas, M. (1979). *Holographic Interferometry*. Berlin: Springer-Verlag.

Smith, H. M. (1969). *Principles of Holography*. New York: Wiley-Interscience (2nd edn. 1975).

Smith, H. M. ed. (1977). *Holographic Recording Materials*. Berlin: Springer-Verlag.

Solymar, L. & Cooke, D. J. (1981). *Volume Holography and Volume Gratings*. New York: Academic Press.

Soroko, L. M. (1980). *Holography and Coherent Optics*. (Transl. A. Tybulewicz). New York: Plenum Press.

Stroke, G. W. (1966). *An Introduction to Coherent Optics and Holography*. New York: Academic Press. (2nd edition, 1969).

Vest, C. M. (1979). *Holographic Interferometry*. New York: John Wiley.

von Bally, G. ed. (1979). *Holography in Medicine and Biology*. Berlin: Springer-Verlag.

Wenyon, M. (1978). *Understanding Holography*. Newton Abbot: David & Charles.

Yaroslavskii, L. P. & Merzlyakov, N. S. (1980). *Methods of Digital Holography*. New York: Consultants Bureau, Plenum Publishing Co.

Yu, F. T. S. (1973). *Introduction to Diffraction, Information Processing and Holography*. Cambridge: The MIT Press.

References

Abramson, N. (1969). The holo-diagram: a practical device for making and evaluating holograms. *Applied Optics*, **8**, 1235–40.

Abramson, N. (1970*a*). The holo-diagram II. A practical device for information retrieval in hologram interferometry. *Applied Optics*, **9**, 97–101.

Abramson, N. (1970*b*). The holo-diagram III. A practical device for predicting fringe patterns in hologram interferometry. *Applied Optics*, **9**, 2311–20.

Abramson, N. (1971). The holo-diagram IV. A practical device for simulating fringe patterns in hologram interferometry. *Applied Optics*, **10**, 2155–61.

Abramson, N. (1972). The holo-diagram V. A device for practical interpreting of hologram fringes. *Applied Optics*, **11**, 1143–7.

Abramson, N. (1974). Sandwich hologram interferometry: a new dimension in holographic comparison. *Applied Optics*, **13**, 2019–25.

Abramson, N. (1975). Sandwich hologram interferometry. 2. Some practical calculations. *Applied Optics*, **14**, 981–4.

Abramson, N. (1976*a*). Sandwich hologram interferometry. 3: Contouring. *Applied Optics*, **15**, 200–5.

Abramson, N. (1976*b*). Holographic contouring by translation. *Applied Optics*, **15**, 1018–22.

Abramson, N. (1977). Sandwich hologram interferometry. 4: Holographic studies of two milling machines. *Applied Optics*, **16**, 2521–31.

Abramson, N. H. & Bjelkhagen, H. (1973). Industrial holographic measurements. *Applied Optics*, **11**, 2792–6.

Abramson, N. H. & Bjelkhagen, H. (1978). Pulsed sandwich holography. 2. Practical application. *Applied Optics*, **17**, 187–91.

Abramson, N. & Bjelkhagen, H. (1979) Sandwich hologram interferometry. 5: Measurement of in-plane displacement and compensation for rigid body motion. *Applied Optics*, **18**, 2870–80.

Abramson, N., Bjelkhagen, H. & Skande, P. (1979). Sandwich holography for storing information interferometrically with a high degree of accuracy. *Applied Optics*, **18**, 2017–21.

Ågren, C.-H. & Stetson, K. A. (1972). Measuring the resonances of treble viol plates by hologram interferometry and designing an improved instrument. *Journal of the Acoustical Society of America*, **51**, 1971–83.

Aleksandrov, E. G. & Bonch-Bruevich, A. M. (1967). Investigation of surface strains by the hologram technique. *Soviet Physics: Technical Physics*, **12**, 258–65.

Aleksoff, C. C. (1971). Temporally modulated holography. *Applied Optics*, **10**, 1329–41.

Alferness, R. (1975*a*). Analysis of optical propagation in thick holographic gratings. *Applied Physics*, **7**, 29–33.

273

Alferness, R. (1975b). Equivalence of the thin-grating decomposition and coupled wave analysis of thick holographic gratings. *Optics Communications*, **15**, 209–12.

Alferness, R. (1976). Analysis of propagation at the second-order Bragg angle of a thick holographic grating. *Journal of the Optical Society of America*, **66**, 353–62.

Alferness, R. & Case, S. K. (1975). Coupling in doubly-exposed, thick holographic gratings. *Journal of the Optical Society of America*, **65**, 730–9.

Altman, J. H. (1966). Pure relief images in type 649 F plates. *Applied Optics*, **5**, 1689–90.

Amadesi, S., Gori, F., Grella, R. & Guattari, G. (1974). Holographic methods for painting diagnostics. *Applied Optics*, **13**, 2009–13.

Anderson, L. K. (1968). Holographic optical memory for bulk data storage. *Bell Laboratories Record*, **46**, 318–25.

Andreev, R. B., Vorzobova, N. D., Kalintsev, A. G. & Staselko, D. I. (1980). Image pulsed holography recording in the green region of the spectrum. *Optics & Spectroscopy*, **49**, 514–15.

ANSI (1980). *American National Standard for the Safe Use of Lasers*, **Z 136**. 1–1980. New York: American Standards Institute.

Ansley, D. A. (1970). Techniques for pulsed laser holography of people. *Applied Optics*, **9**, 815–21.

Aoyagi, Y. & Namba, S. (1976). Blazing of holographic grating by ion etching technique. *Japanese Journal of Applied Physics*, **15**, 721–2.

Aoyagi, Y., Sano, K. & Namba, S. (1979). High spectroscopic qualities in blazed ion-etched holographic gratings. *Optics Communications*, **29**, 253–5.

Archbold, E. & Ennos, A. E. (1968). Observation of surface vibration modes by stroboscopic hologram interferometry. *Nature*, **217**, 942–3.

Aristov, V. V., Shekhtman, V. Sh. & Timofeev, V. B. (1969). The Borrmann effect and extinction in holography. *Physics Letters*, **28A**, 700–1.

Armstrong, W. T. & Forman, P. R. (1977). Double-pulsed time differential holographic interferometry. *Applied Optics*, **16**, 229–32.

Assa, A. & Betser, A. A. (1974). The application of holographic multiplexing to record separate isopachic- and ischromatic-fringe patterns. *Experimental Mechanics*, **14**, 502–4.

Balasubramanian, N. (1975). Holographic applications in photogrammetry. *Optical Engineering*, **14**, 448–52.

Ballard, G. S. (1968). Double exposure holographic interferometry with separate reference beams. *Journal of Applied Physics*, **39**, 4846–8.

Bar-Joseph, I., Hardy, A., Katzir, Y. & Silberberg, Y. (1981). Low-power phase-conjugate interferometry. *Optics Letters*, **6**, 414–16.

Bartolini, R. A. (1972). Improved development for holograms recorded in photoresist. *Applied Optics*, **11**, 1275–6.

Bartolini, R. A. (1974). Characteristics of relief phase holograms recorded in photoresists. *Applied Optics*, **13**, 129–39.

Bartolini, R. A. (1977a). Photoresists. In *Holographic Recording Materials*, ed. H. M. Smith, pp. 209–27. Berlin: Springer Verlag.

Bartolini, R. A. (1977b). Optical recording media review. In *Optical Storage Materials & Methods*, Proceedings of the SPIE, vol. 123, ed. L. Beiser & D. Chen, pp. 2–9. Bellingham: SPIE.

Bartolini, R., Hannan, W., Karlsons, D. & Lurie, M. (1970). Embossed hologram motion pictures for television playback. *Applied Optics*, **9**, 2283–90.

Bates, H. E. (1973). Burst-mode frequency-doubled YAG:Nd^{3+} laser for time-sequenced high speed photography and holography. *Applied Optics*, **12**, 1172–8.

Beesley, M. J., Castledine, J. G. & Cooper, D. P. (1969). Sensitivity of resist coated silicon slices to argon laser wavelengths. *Electronics Letters*, **5**, 257–8.

Beesley, M. J., Foster, H. & Hambleton, K. G. (1968). Holographic projection of microcircuit patterns. *Electronics Letters*, **4**, 49–50.

Belvaux, Y. (1967). Hologram duplication. *Annales de Radioelectricité*, **22**, 105–8.

Belvaux, Y. (1975). Influence de divers paramètres d'enregistrement lors de la restitution d'un hologramme. *Nouvelle Revue d'Optique*, **6**, 137–47.

Benlarbi, B., Cooke, D. J. & Solymar, L. (1980). Higher order modes in thick phase gratings. *Optica Acta*, **27**, 885–95.

Bennett, S. J. (1976). Achromatic combinations of hologram optical elements. *Applied Optics*, **15**, 542–5.

Benton, S. A. (1969). Hologram reconstructions with extended incoherent sources. *Journal of the Optical Society of America*, **59**, 1545–6.

Benton, S. A. (1971). Granularity effects in phase holograms. *Journal of the Optical Society of America*, **61**, 649.

Benton, S. A. (1975). Holographic displays – a review. *Optical Engineering*, **14**, 402–7.

Benton, S. A. (1977). White light transmission/reflection holographic imaging. In *Applications of Holography & Optical Data Processing*, ed. E. Marom, A. A. Friesem & E. Wiener-Avnear, pp. 401–9. Oxford: The Pergamon Press.

Benton, S. A. (1978). Achromatic images from white-light transmission holograms. *Journal of the Optical Society of America*, **68**, 1441.

Benton, S. A. (1980). Holographic displays: 1975–1980. *Optical Engineering*, **19**, 686–90.

Benton, S. A., Mingace, Jr, H. S. & Walter, W. R. (1979). In *Optics and Photonics Applied to 3-Dimensional Imagery*, Proceedings of the SPIE, vol. 212, ed. M. Grosmann & P. Meyrueis, pp. 2–7. Bellingham: SPIE.

Biedermann, K. (1969). A function characterizing photographic film that directly relates to brightness of holographic images. *Optik*, **28**, 160–76.

Biedermann, K. (1970). The scattered flux spectrum of photographic materials for holography. *Optik*, **31**, 367–89.

Biedermann, K. & Holmgren, O. (1977). Large-size distortion-free computer-generated holograms in photoresist. *Applied Optics*, **16**, 2014–16.

Biedermann, K. & Johansson, S. (1972). Evaluation of the modulation transfer function of photographic emulsions by means of a multiple-sine-slit microdensitometer. *Optik*, **35**, 391–403.

Biedermann, K. & Johansson, S. (1975). A universal instrument for the evaluation of the MTF and other recording parameters of photographic materials. *Journal of Physics E: Scientific Instruments*, **8**, 751–7.

Biedermann, K. & Molin, N.-E. (1970). Combining hypersensitization and *in situ* processing for time-average observation in real-time hologram interferometry. *Journal of Physics E: Scientific Instruments*, **3**, 669–80.

Birch, K. G. & Green, F. J. (1972). The application of computer-generated holograms to testing optical elements. *Journal of Physics D: Applied Physics*, **5**, 1982–92.

Bjelkhagen, H. (1974). Holographic time-average vibration study of a structure dynamic model of an airplane fin. *Optics & Laser Technology*, **6**, 117–23.

Bjelkhagen, H. I. (1977a). Experiences with large-scale reflection and transmission holograms. In *Three-Dimensional Imaging*, Proceedings of the SPIE, vol. 120, ed. S. A. Benton, pp. 122–6. Redondo Beach: SPIE.

Bjelkhagen, H. (1977b). Pulsed sandwich holography. *Applied Optics*, **16**, 1727–31.

Bloom, A. L. (1968). *Gas Lasers*. New York: John Wiley.

Boone, P. M. (1975). Use of reflection holograms in holographic interferometry and speckle correlation for measurement of surface displacement. *Optica Acta*, **22**, 579–89.

Boone, P. & Verbiest, R. (1969). Application of hologram interferometry to plate deformation and translation measurements. *Optica Acta*, **16**, 555–67.

Booth, B. L. (1975). Photopolymer material for holography. *Applied Optics*, **14**, 593–601.

Booth, B. L. (1977). Photopolymer laser recording materials. *Journal of Applied Photographic Engineering*, **3**, 24–30.

Born, M. & Wolf, E. (1980). *Principles of Optics*. New York: Pergamon Press.

Bracewell, R. (1965). *The Fourier Transform & its Applications*. New York: McGraw-Hill (2nd edition: 1978).

Bragg, W. L. (1939). A new type of 'X-ray microscope'. *Nature*, **143**, 678.

Bragg, W. L. (1942). The X-ray microscope. *Nature*, **149**, 470–71.

Brandes, R. G., Francois, E. E. & Shankoff, T. A. (1969). Preparation of dichromated gelatin films for holography. *Applied Optics*, **8**, 2346–8.

Breidne, M., Johansson, S., Nilsson, L.-E. & Åhlen, H. (1979). Blazed holographic gratings. *Optica Acta*, **26**, 1427–41.

Briers, J. D. (1976). The interpretation of holographic interferograms. *Optics & Quantum Electronics*, **8**, 469–501.

Brooks, R. E., Heflinger, L. O. & Wuerker, R. F. (1965). Interferometry with a holographically reconstructed comparison beam. *Applied Physics Letters*, **7**, 248–9.

Brown, B. R. & Lohman, A. W. (1966). Complex spatial filtering with binary masks. *Applied Optics*, **5**, 967–9.

Brown, B. R. & Lohmann, A. W. (1969). Computer generated binary holograms. *IBM Journal of Research & Development*, **13**, 160–7.

Brumm, D. B. (1967). Double images in copy holograms. *Applied Optics*, **6**, 588–9.

Bryngdahl, O. (1967). Polarizing holography, *Journal of the Optical Society of America*, **57**, 545–7.

Bryngdahl, O. (1969a). Longitudinally reversed shearing interferometry. *Journal of the Optical Society of America*, **59**, 142–6.

Bryngdahl, O. (1969b). Holography with evanescent waves. *Journal of the Optical Society of America*, **59**, 1645–50.

Bryngdahl, O. (1973). Evanescent waves in optical imaging. In *Progress in Optics*, vol. 11, ed. E. Wolf, pp. 169–221. Amsterdam: North-Holland.

Bryngdahl, O. (1974a). Optical map transformations. *Optics Communications*, **10**, 164–8.

Bryngdahl, O. (1974b). Geometrical transformations in optics. *Journal of the Optical Society of America*, **64**, 1092–9.

Bryngdahl, O. (1975). Computer generated holograms as generalized optical components. *Optical Engineering*, **14**, 426–35.

Bryngdahl, O. & Lohmann, A. W. (1968a). Interferograms are image holograms. *Journal of the Optical Society of America*, **58**, 141–2.

Bryngdahl, O. & Lohmann, A. (1968*b*). One dimensional holography with spatially incoherent light. *Journal of the Optical Society of America*, **58**, 625–8.

Bryngdahl, O. & Lohmann, A. (1968*c*). Non-linear effects in holography. *Journal of the Optical Society of America*, **58**, 1325–34.

Bryngdahl, O. & Lohmann, A. (1970*a*). Variable magnification in incoherent holography. *Applied Optics*, **9**, 231–2.

Bryngdahl, O. & Lohmann, A. (1970*b*). Holography in white light. *Journal of the Optical Society of America*, **60**, 281–3.

BSI (1983). *British Standard Guide on Protection of Personnel Against Hazards From Laser Radiation*, BS 4803: 1983. London: British Standards Institution.

Burch, J. J. (1967). A computer algorithm for the synthesis of spatial frequency filters. *Proceedings of the IEEE*, **55**, 599–601.

Burch, J. M. (1965). The application of lasers in production engineering. *The Production Engineer*, **44**, 431–42.

Burch, J. M. & Palmer, D. A. (1961). Interferometric methods for the photographic production of large gratings. *Optica Acta*, **8**, 73–80.

Burckhardt, C. B. (1966*a*). Diffraction of a plane wave at a sinusoidally stratified dielectric grating. *Journal of the Optical Society of America*, **56**, 1502–9.

Burckhardt, C. B. (1966*b*). Display of holograms in white light. *Bell Systems Technical Journal*, **45**, 1841-4.

Burckhardt, C. B. (1967). Efficiency of a dielectric grating. *Journal of the Optical Society of America*, **57**, 601–3.

Burckhardt, C. B. (1970). A simplification of Lee's method of generating holograms by computer. *Applied Optics*, **9**, 1949.

Burke, W. J., Staebler, D. L., Phillips, W. & Alphonse, G. A. (1978). Volume phase holographic storage in ferroelectric crystals. *Optical Engineering*, **17**, 308–16.

Buschmann, H. T. (1971). The production of low noise, bright, phase holograms by bleaching. *Optik*, **34**, 240–53.

Buschmann, H. T. (1972). The wavelength dependence of the transfer properties of photographic materials for holography. In *Optical & Acoustical Holography*, ed. E. Camatini, pp. 151–72. New York: The Plenum Press.

Buschmann, H. T. & Metz, H. J. (1971). Die wellenlängenabhängigkeit der übertragungseigenschaften photographischer materialen für die holographie. *Optics Communications*, **2**, 373–6.

Butusov, M. M. & Ioffe, A. I. (1976). Investigation of parameters of holographic periodic multiple-imaging structures. *Soviet Journal of Quantum Electronics*, **6**, 519–21.

Casasent, D. ed. (1978). *Optical Data Processing*. Topics in Applied Physics, vol. 23. Berlin: Springer-Verlag.

Casasent, D. & Caimi, F. (1977). Photodichroic crystals for coherent optical data processing. *Optics & Laser Technology*, **9**, 63–8.

Case, S. K. (1975). Coupled wave theory for multiply exposed thick holographic gratings. *Journal of the Optical Society of America*, **65**, 724–9.

Case, S. K. & Alferness, R. (1976). Index modulation and spatial harmonic generation in dichromated gelatin films. *Applied Physics*, **10**, 41–51.

Case, S. K., Haugen, P. R. & Løkberg, O. J. (1981). Multifacet holographic optical elements for wavefront transformations. *Applied Optics*, **20**, 2670–5.

Cathey Jr, W. T. (1965). Three dimensional wavefront reconstruction using a phase hologram. *Journal of the Optical Society of America*, **55**, 457.

Caulfield, H. J. (1972). Multiplexing double-exposure holographic interferograms. *Applied Optics*, **11**, 2711–12.

Caulfield, H. J. & Beyen, W. J. (1967). Birefringent beam splitting for holography. *Review of Scientific Instruments*, **38**, 977–8.

Cha, S. & Vest, C. M. (1979). Interferometry and reconstruction of strongly refracting asymmetric refractive-index fields. *Optics Letters*, **4**, 311–13.

Cha, S. & Vest, C. M. (1981). Tomographic reconstruction of strongly refracting fields and its application to interferometric measurement of boundary layers. *Applied Optics*, **20**, 2787–94.

Champagne, E. B. (1967). Non-paraxial imaging, magnification and aberration properties in holography. *Journal of the Optical Society of America*, **57**, 51–5.

Champagne, E. B. & Massey, N. G. (1969). Resolution in holography. *Applied Optics*, **8**, 1879–85.

Chang, B. J. (1973). Holography with non-coherent light. *Optics Communications*, **9**, 357–9.

Chang, B. J. (1976). Post-processing of developed dichromated gelatin holograms. *Optics Communications*, **17**, 270–1.

Chang, B. J. (1979). Dichromated gelatin as a holographic storage medium. In *Optical Information Storage*, Proceedings of the SPIE, vol. 177, ed. K. G. Leib, pp. 71–81. Bellingham: SPIE.

Chang, B. J. & Leith, E. N. (1979). Space-invariant multiple-grating interferometers in holography. *Journal of the Optical Society of America*, **69**, 689–96.

Chang, B. J. & Leonard, C. D. (1979). Dichromated gelatin for the fabrication of holographic optical elements. *Applied Optics*, **18**, 2407–17.

Chang, B. J. & Winick, K. (1980). Silver-halide gelatin holograms. In *Recent Advances in Holography*, Proceedings of the SPIE, vol. 215, ed. T. C. Lee & P. N. Tamura, pp. 172–7. Bellingham: SPIE.

Chang, M. & George, N. (1970). Holographic dielectric grating: theory and practice. *Applied Optics*, **9**, 713–19.

Chatelain, B. (1973). Holographic photo-elasticity: independent observation of the isochromatic and isopachic fringes for a single model subjected to only one process. *Optics & Laser Technology*, **5**, 201–4.

Chau, H. H. M. (1968). Holographic interferometer for isopachic stress analysis. *Review of Scientific Instruments*, **39**, 1789–92.

Chau, H. M. (1970). A full view holographic system. *Applied Optics*, **9**, 1479–80.

Chen, H. (1979). Astigmatic one-step rainbow hologram process. *Applied Optics*, **18**, 3728–30.

Chen, H., Tai, A. & Yu, F. T. S. (1978). Generation of colour images with one-step rainbow holograms. *Applied Optics*, **17**, 1490–1.

Chen, H. & Yu. F. T. S. (1978). One-step rainbow hologram. *Optics Letters*, **2**, 85–7.

Chomat, M. & Miler, M. (1973). Application of holography to the analysis of mechanical vibration in electronic components. *TESLA Electronics*, **3**, 83–93.

Chu, D. C., Fienup, J. R. & Goodman, J. W. (1973). Multi-emulsion, on-axis, computer generated holograms. *Applied Optics*, **12**, 1386–8.

Chu, R. S. & Tamir, T. (1970). Guided-wave theory of light diffraction by acoustic microwaves. *IEEE Transactions on Microwave Theory & Techniques*, MTT-18, 486–504.

Cindrich, I. (1967). Image scanning by rotation of a hologram. *Applied Optics*, **6**, 1531–4.

Close, D. H. (1975). Holographic optical elements. *Optical Engineering*, **14**, 408–19.

Cochran, G. (1966). New method of making Fresnel transforms with incoherent light. *Journal of the Optical Society of America*, **56**, 1513–17.

Cochran, W. T., Cooley, J. W., Favin, D. L., Helms, H. D., Kaenel, R. A., Lang, W. W., Maling Jr, G. C., Nelson, D. E., Rader, C. M., & Welch, P. D. (1967). What is the fast Fourier transform? *Proc. IEEE*, **55**, 1664–74.

Colburn, W. S. & Dubow, J. B. (1973). Photoplastic recording materials. Technical Report AFAL-TR-73-255. Ann Arbor: Harris Electro-Optics Center of Radiation.

Coleman, D. J. & Magariños, J. (1981). Controlled shifting of the spectral response of reflection holograms. *Applied Optics*, **20**, 2600–1.

Collier, R. J., Burckhardt, C. B. & Lin, L. H. (1971). *Optical Holography*. New York, Academic Press.

Collier, R. J., Doherty, E. T. & Pennington, K. S. (1965). Application of moire techniques to holography. *Applied Physics Letters*, **7**, 223–5.

Collier, R. J. & Pennington, K. S. (1966). Ghost imaging in holograms formed in the near field. *Applied Physics Letters*, **8**, 44–6.

Collier, R. J. & Pennington, K. S. (1967). Multicolor imaging from holograms formed on two-dimensional media. *Applied Optics*, **6**, 1091–5.

Collins, L. F. (1968). Difference holography. *Applied Optics*, **7**, 203–5.

Credelle, T. L. & Spong, F. W. (1972). Thermoplastic media for holographic recording. *RCA Review*, **33**, 206–26.

Curran, R. K. & Shankoff, T. A. (1970). The mechanism of hologram formation in dichromated gelatin. *Applied Optics*, **9**, 1651–7.

Cutrona, L. J. (1960). Optical data processing and filtering systems. *IEEE Transactions on Information Theory*, IT-6, 386–400.

Dainty, J. C. & Welford, W. T. (1971). Reduction of speckle in image plane hologram reconstruction by moving pupils. *Optics Communications*, **3**, 289–94.

Dallas, W. J. (1971a). Phase quantization – a compact derivation. *Applied Optics*, **10**, 673–4.

Dallas, W. J. (1971b). Phase quantization in holograms – a few illustrations. *Applied Optics*, **10**, 674–6.

Dallas, W. J. (1980). Computer-generated holograms. In *The Computer in Optical Research*, Topics in Applied Physics, vol. 41, ed. B. R. Frieden, pp. 291–366. Berlin: Springer-Verlag.

Dallas, W. J. & Lohmann, A. W. (1975). Deciphering vibration holograms. *Optics Communications*, **13**, 134–7.

Dändliker, R. (1980). Heterodyne holographic interferometry. In *Progress in Optics*, vol. 17, ed. E. Wolf, pp. 1–84. Amsterdam: North-Holland.

Dändliker, R., Ineichen, B. & Mottier, F. M. (1973). High resolution hologram interferometry by electronic phase measurement. *Optics Communications*, **9**, 412–16.

Dändliker, R., Thalmann, R. & Willemin, J.-F. (1982). Fringe interpolation by two-reference-beam holographic interferometry: reducing sensitivity to hologram misalignment. *Optics Communications*, **42**, 301–6.

D'Auria, L., Huignard, J. P., Slezak, C. & Spitz, E. (1974). Experimental holographic read–write memory using 3-D storage. *Applied Optics*, **13**, 808–18.

De Bitteto, D. J. (1966). White light viewing of surface holograms by simple dispersion compensation. *Applied Physics Letters*, **9**, 417–18.

De Bitteto, D. J. (1969). Holographic panoramic stereograms synthesized from white light recordings. *Applied Optics*, **8**, 1740–41.

De Bitteto, D. J. (1970). A front lighted 3-D holographic movie. *Applied Optics*, **9**, 498–9.

Deen, L. M., Walkup, J. F. & Hagler, M. O. (1975). Representations of space-variant optical systems using volume holograms. *Applied Optics*, **14**, 2438–46.

De Mattia, P. & Fossati-Bellani, V. (1978). Holographic contouring by displacing the object and the illumination beam. *Optics Communications*, **26**, 17–21.

Denisyuk, Yu. N. (1962). Photographic reconstruction of the optical properties of an object in its own scattered radiation field. *Soviet Physics – Doklady*, **7**, 543–5.

Denisyuk, Yu. N. (1963). On the reproduction of the optical properties of an object by the wave field of its scattered radiation. *Optics & Spectroscopy*, **15**, 279–84.

Denisyuk, Yu. N. (1965). On the reproduction of the optical properties of an object by the wave field of its scattered radiation. II. *Optics & Spectroscopy*, **18**, 152–7.

Denisyuk, Yu. N. (1980). Holography and its prospects. *Journal of Applied Spectroscopy*, **33**, 901–15.

Dietrich, H. F., Raine, R. J. & O'Brien, R. N. (1976). A 5-minute monobath for Kodak 649-F plates used in holography and holographic interferometry. *Journal of Photographic Science*, **24**, 120–3.

Dubas, M. & Schumann, W. (1974). Sur la détermination holographique de l'état de déformation à la surface d'un corps non-transparent. *Optica Acta*, **21**, 547–62.

Dubas, M. & Schumann, W. (1975). On direct measurement of strain and rotation in holographic interferometry using the line of complete localization. *Optica Acta*, **22**, 807–19.

Dubas, M. & Schumann, W. (1977). Contribution à l'étude théorique des images et des franges produites par deux hologrammes en sandwich. *Optica Acta*, **24**, 1193–209.

Duncan Jr, R. C. & Staebler, D. L. (1977). Inorganic photochromic materials. In *Holographic Recording Materials*, Topics in Applied Physics, vol. 20, ed. H. M. Smith, pp. 133–60. Berlin: Springer-Verlag.

Ebbeni, J., Coenen, J. & Hermanne, A. (1976). New analysis of holophotoelastic patterns and their application. *Journal of Strain Analysis*, **11**, 11–17.

Elias, P., Grey, D. S. & Robinson, D. Z. (1952). Fourier treatment of optical processes. *Journal of the Optical Society of America*, **42**, 127–34.

El-Sum, H. M. A. & Kirkpatrick, P. (1952). Microscopy by reconstructed wavefronts. *Physical Review*, **85**, 763.

Ennos, A. E. (1968). Measurement of in-plane surface strain by hologram interferometry. *Journal of Physics E: Scientific Instruments*, **1**, 731–4.

Erf, R. K. (1974). *Holographic Non-Destructive Testing*. New York: Academic Press.

Eschler, H. (1975). Multifrequency acousto-optic page composers for holographic data storage. *Optics Communications*, **13**, 148–53.

Ewan, B. C. R. (1979). Particle velocity distribution measurement by holography. *Applied Optics*, **18**, 3156–60.

Fainman, Y., Lenz, E. & Shamir, J. (1981). Contouring by phase conjugation. *Applied Optics*, **20**, 158–63.

Fairchild, R. C. & Fienup, J. R. (1982). Computer-originated aspheric holographic optical elements. *Optical Engineering*, **21**, 133–40.

Faulde, M., Fercher, A. F., Torge, R. & Wilson, R. N. (1973). Optical testing by means of synthetic holograms and partial lens compensation. *Optics Communications*, **7**, 363–5.

Favre, H. (1929). Sur une nouvelle méthode optique de détermination de tensions intérieures. *Revue d'Optique*, **8**, 193–213, 241–61, 289–307.

Fercher, A. F. (1976). Computer generated holograms for testing optical elements: error analysis and error compensation. *Optica Acta*, **23**, 347–65.

Forshaw, M. R. B. (1973). The imaging properties and aberrations of thick transmission holograms. *Optica Acta*, **20**, 669–86.

Forshaw, M. R. B. (1974). Diffraction of a narrow laser beam by a thick hologram: experimental results. *Optics Communications*, **12**, 279–81.

Forshaw, M. R. B. (1975). Explanation of the diffraction fine structure in overexposed thick holograms. *Optics Communications*, **15**, 218–21.

Fourney, M. E. (1968). Application of holography to photoelasticity. *Experimental Mechanics*, **8**, 33–8.

Fourney, M. E. & Mate, K. V. (1970). Further applications of holography to photoelasticity. *Experimental Mechanics*, **10**, 177–86.

Fournier, J.-M., Tribillon, G. & Viénot, J.-C. (1977). Recording of large size holograms in photographic emulsion: image reconstruction. In *Three-Dimensional Imaging*, Proceedings of the SPIE, vol. 120, ed. S. A. Benton, pp. 116–21. Redondo Beach: SPIE.

Friesem, A. A. & Federowicz, R. J. (1966). Recent advances in multicolor wavefront reconstruction. *Applied Optics*, **5**, 1085–6.

Friesem, A. A. & Federowicz, R. J. (1967). Multicolor wavefront reconstruction. *Applied Optics*, **6**, 529–36.

Friesem, A. A. & Levy, U. (1976). Fringe formation in two-wavelength contour holography. *Applied Optics*, **15**, 3009–20.

Friesem, A. A. & Walker, J. L. (1970). Thick absorption recording media in holography. *Applied Optics*, **9**, 201–14.

Friesem, A. A. & Zelenka, J. S. (1967). Effects of film non-linearities in holography. *Applied Optics*, **6**, 1755–9.

Gabor, D. (1948). A new microscopic principle. *Nature*, **161**, 777–8.

Gabor, D. (1949). Microscopy by reconstructed wavefronts. *Proceedings of the Royal Society A*, **197**, 454–87.

Gabor, D. (1951). Microscopy by reconstructed wavefronts. II. *Proceedings of the Physical Society (Lond.) B*, **64**, 449–69.

Gabor, D. (1969). Associative holographic memories. *IBM Journal of Research & Development*, **13**, 156–9.

Gabor, D. (1970). Laser speckle and its elimination. *IBM Journal of Research & Development*, **14**, 509–14.

Gale, M. T. & Knop, K. (1976). Color-encoded focused image holograms. *Applied Optics*, **15**, 2189–98.

Gale, M. T., Knop, K. & Russell, J. P. (1975). A colour micro-storage and display system using focused image holograms. *Optics & Laser Technology*, **7**, 234–6.

Gara, A. D., Majkowski, R. F. & Stapleton, T. T. (1973). Holographic system for automatic surface mapping. *Applied Optics*, **12**, 2172–9.

Gåsvik, K. (1975). Holographic reconstruction of the state of polarization. *Optica Acta*, **22**, 189–206.

Gates, J. W. C. (1968). Holographic phase recording by interference between reconstructed wavefronts from separate holograms. *Nature*, **220**, 473–4.

Gates, J. W. C., Hall, R. G. N. & Ross, I. N. (1970). Holographic recording using frequency-doubled radiation at 530 nm. *Journal of Physics E: Scientific Instruments*, **3**, 89–94.

Gates, J. W. C., Hall, R. G. N. & Ross, I. N. (1972). Holographic interferometry of impact-loaded objects using a double-pulse laser. *Optics & Laser Technology*, **4**, 72–5.

Gaylord, T. K. & Moharam, M. G. (1982). Planar dielectric grating diffraction theories. *Applied Physics B*, **28**, 1–14.

George, N. (1970). Full view holograms. *Optics Communications*, **1**, 457–9.

George, N. & Jain, A. (1973). Speckle reduction using multiple tones of illumination. *Applied Optics*, **12**, 1202–12.

Gerasimova, S. A. & Zakharchenko, V. M. (1981). Holographic processor for associative information retrieval. *Soviet Journal of Optical Technology*, **48**, 404–6.

Gerritsen, H. J., Hannan, W. J. & Ramberg, E. G. (1968). Elimination of speckle noise in holograms with redundancy. *Applied Optics*, **7**, 2301–11.

Ghandeharian, H. & Boerner, W. M. (1977). Autocorrelation of transmittance of holograms made of diffuse objects. *Optica Acta*, **24**, 1087–97.

Ghandeharian, H. & Boerner, W. M. (1978). Degradation of holographic images due to depolarization of reflected light. *Journal of the Optical Society of America*, **68**, 931–4.

Glaser, I. (1973). Anamorphic imagery in holographic stereograms. *Optics Communications*, **7**, 323–6.

Glaser, I. & Friesem, A. A. (1977). Imaging properties of holographic stereograms. In *Three-Dimensional Imaging*, Proceedings of the SPIE, vol. 120, ed. S. A. Benton, pp. 150–62. Redondo Beach: SPIE.

Glass, A. M. (1978). The photorefractive effect. *Optical Engineering*, **17**, 470–9.

Golbach, H. (1973). Reduction of speckle in holographic reflected-light microscopy. *Optik*, **37**, 45–9.

Goldberg, J. L. (1975). A holographic interferometer for the measurement of the vector displacement of a slowly deforming rough surface. *Japanese Journal of Applied Physics*, **14** (Supplement 14-1), 253–8.

Goodman, J. W. (1967). Film grain noise in wavefront reconstruction imaging. *Journal of the Optical Society of America*, **57**, 493–502.

Goodman, J. W. (1968). *Introduction to Fourier Optics*. New York: McGraw-Hill.

Goodman, J. W. (1975). Statistical properties of laser speckle patterns. In *Laser Speckle & Related Phenomena*, Topics in Applied Physics, vol. 9, ed. J. C. Dainty, pp. 9–75. Berlin: Springer-Verlag.

Goodman, J. W. (1981). Linear space-variant optical data processing. In *Optical Information Processing: Fundamentals*, Topics in Applied Physics, vol. 48, ed. S. H. Lee, pp. 235–60. Berlin: Springer-Verlag.

Goodman, J. W., Huntley, W. H., Jackson, D. W. & Lehmann, M. (1966). Wavefront reconstruction imaging through random media. *Applied Physics Letters*, **8**, 311–13.

Goodman, J. W. & Knight, G. R. (1968). Effects of film nonlinearities on wavefront-reconstruction images of diffuse objects. *Journal of the Optical Society of America*, **58**, 1276–83.

Goodman, J. W. & Silvestri, A. M. (1970). Some effects of Fourier domain phase quantization. *IBM Journal of Research & Development*, **14**, 478–84.

Graube, A. (1973). Holograms recorded with red light in dye sensitized dichromated gelatin. *Optics Communications*, **8**, 251–3.

Graube, A. (1974). Advances in bleaching methods for photographically recorded holograms. *Applied Optics*, **13**, 2942–6.

Graver, W. R., Gladden, J. W. & Estes, J. W. (1980). Phase holograms formed by silver halide (sensitized) gelatin processing. *Applied Optics*, **19**, 1529–36.

Greenaway, D. L. (1980). Cards and card-readers for voucher and access control systems. *Landis & Gyr Review*, **27**, 20–5.

Greguss, P. (1975). *Holography in Medicine*. London: IPC Press.

Greguss, P. (1976). Holographic interferometry in biomedical sciences. *Optics & Laser Technology*, **8**, 153–9.

Groh, G. (1968). Multiple imaging by means of point holograms. *Applied Optics*, **7**, 1643–4.

Groh, G. & Kock, M. (1970). 3-D display of X-ray images by means of holography. *Applied Optics*, **9**, 775–7.

Gupta, P. C. & Singh, K. (1975a). Characteristic fringe function for time-average holography of periodic nonsinusoidal vibrations. *Applied Optics*, **14**, 129–33.

Gupta, P. C. & Singh, K. (1975b). Time-average hologram interferometry of periodic, non-cosinusoidal vibrations. *Applied Physics*, **6**, 233–40.

Gupta, P. C. & Singh, K. (1976). Hologram interferometry of vibrations represented by the square of a Jacobian elliptic function. *Nouvelle Revue d'Optique*, **7**, 95–100.

Guther, R. & Kusch, S. (1974). Ein beitrag zum Intermodulationsrauschen in der Volumenholographie. *Experimentelle Technik der Physik*, **22**, 119–41.

Haig, N. D. (1973). Three dimensional holograms by rotational multiplexing of two-dimensional films. *Applied Optics*, **12**, 419–20.

Haines, K. A. & Hildebrand, B. P. (1965). Contour generation by wavefront reconstruction. *Physics Letters*, **19**, 10–11.

Haines, K. A. & Hildebrand, B. P. (1966). Surface-deformation measurement using the wavefront reconstruction technique. *Applied Optics*, **5**, 595–602.

Hamasaki, J. (1968). Signal-to-noise ratios for hologram images of subjects in strong incoherent light. *Applied Optics*, **7**, 1613–20.

Hannan, W. J., Flory, R. E., Lurie, M. & Ryan, R. J. (1973). Holotape: a low-cost prerecorded television system using holographic storage. *Journal of the Society of Motion Picture and Television Engineers*, **82**, 905–15.

Hariharan, P. (1971). Reversal processing technique for phase holograms. *Optics Communications*, **3**, 119–21.

Hariharan, P. (1972). Bleached reflection holograms. *Optics Communications*, **6**, 377–9.

Hariharan, P. (1973). Application of holographic subtraction to time-average hologram interferometry of vibrating objects. *Applied Optics*, **12**, 143–6.

Hariharan, P. (1976a). Longitudinal distortion in images reconstructed by reflection holograms. *Optics Communications*, **17**, 52–4.

Hariharan, P. (1976b). Comment on: Sensitivity improvement by step-biasing in holographic interferometry. *Optical Engineering*, **15**, 279.

Hariharan, P. (1977a). Hologram interferometry: identification of the sign of surface displacements. *Optica Acta*, **24**, 989–90.

Hariharan, P. (1977b). Simple full-view rainbow holograms. *Optical Engineering*, **16**, 520–2.

Hariharan, P. (1978). Hologram recording geometry: its influence on image luminance. *Optica Acta*, **25**, 527–30.

Hariharan, P. (1979a). Intermodulation noise in amplitude holograms: the effect of hologram thickness. *Optica Acta*, **26**, 211–15.

Hariharan, P. (1979b). Volume phase reflection holograms: the effect of hologram thickness on image luminance. *Optica Acta*, **26**, 1443–7.

Hariharan, P. (1980a). Improved techniques for multicolour reflection holograms. *Journal of Optics (Paris)*, **11**, 53–5.

Hariharan, P. (1980b). Holographic recording materials: recent developments. *Optical Engineering*, **19**, 636–41.

Hariharan, P. (1980c). Pseudocolour images with volume reflection holograms. *Optics Communications*, **35**, 42–4.

Hariharan, P. (1982). Concentric etalon for single-frequency operation of high-power ion lasers. *Optics Letters*, **7**, 274–5.

Hariharan, P. (1983). Colour holography. In *Progress in Optics*, vol. 20, ed. E. Wolf, pp. 265–324. Amsterdam: North-Holland.

Hariharan, P. & Hegedus, Z. S. (1973). Simple multiplexing technique for double-exposure hologram interferometry. *Optics Communications*, **9**, 152–5.

Hariharan, P. & Hegedus, Z. S. (1974a). Reduction of speckle in coherent imaging by spatial frequency sampling. *Optica Acta*, **21**, 345–56.

Hariharan, P. & Hegedus, Z. S. (1974b). Reduction of speckle in coherent imaging by spatial frequency sampling. II. Random spatial frequency sampling. *Optica Acta*, **21**, 683–95.

Hariharan, P. & Hegedus, Z. S. (1975). Relative phase shift of images reconstructed by phase and amplitude holograms. *Applied Optics*, **14**, 273–4.

Hariharan, P. & Hegedus, Z. S. (1976). Two-hologram interferometry: a simplified sandwich technique. *Applied Optics*, **15**, 848–9.

Hariharan, P., Hegedus, Z. S. & Steel, W. H. (1979). One-step multicolour rainbow holograms with wide angle of view. *Optica Acta*, **26**, 289–91.

Hariharan, P., Kaushik, G. S., & Ramanathan, C. S. (1972). Reduction of scattering in photographic phase holograms. *Optics Communications*, **5**, 59–61.

Hariharan, P., Oreb, B. F. & Brown, N. (1982). A digital phase-measurement system for real-time holographic interferometry. *Optics Communications*, **41**, 393–6.

Hariharan, P., Oreb, B. F. & Brown, N. (1983). Real-time holographic interferometry: a microcomputer system for the measurement of vector displacements. *Applied Optics*, **22**, 876–80.

Hariharan, P. & Ramanathan, C. S. (1971). Suppression of printout effect in photographic phase holograms. *Applied Optics*, **10**, 2197–9.

Hariharan, P., Ramanathan, C. S. & Kaushik, G. S. (1971). Simplified processing technique for photographic phase holograms. *Optics Communications*, **3**, 246–7.

Hariharan, P., Ramanathan, C. S. & Kaushik, G. S. (1973). Monobath processing for holography. *Applied Optics*, **12**, 611–12.

Hariharan, P. & Ramprasad, B. S. (1972). Simplified optical system for holographic subtraction. *Journal of Physics E: Scientific Instruments*, **5**, 976–8.

Hariharan, P. & Ramprasad, B. S. (1973a). Wavefront tilter for double-exposure holographic interferometry. *Journal of Physics E: Scientific Instruments*, **6**, 173–5.

Hariharan, P. & Ramprasad, B. S. (1973*b*). Rapid *in situ* processing for real-time holographic interferometry. *Journal of Physics E: Scientific Instruments*, **6**, 699–701.

Hariharan, P. & Sen, D. (1961). Radial shearing interferometer. *Journal of Scientific Instruments*, **38**, 428–32.

Hariharan, P., Steel, W. H. & Hegedus, Z. S. (1977). Multicolor holographic imaging with a white light source. *Optics Letters*, **1**, 8–9.

Harris Jr, F. S., Sherman, G. C. & Billings, B. H. (1966). Copying holograms. *Applied Optics*, **5**, 665–6.

Haskell, R. E. (1973). Computer-generated binary holograms with minimum quantization errors. *Journal of the Optical Society of America*, **63**, 504.

Haskell, R. E. & Culver, B. C. (1972). New coding technique for computer-generated holograms. *Applied Optics*, **11**, 2712–14.

Heflinger, L. O., Stewart, G. L. & Booth, C. R. (1978). Holographic motion pictures of microscopic plankton. *Applied Optics*, **17**, 951–4.

Henshaw, P. D. & Ezekiel, S. (1974). High resolution holographic contour generation with white light reconstruction. *Optics Communications*, **12**, 39–42.

Hercher, M. (1969). Tunable single mode operation of gas lasers using intracavity tilted etalons. *Applied Optics*, **8**, 1103–6.

Hermann, J. P., Herriau, J. P. & Huignard, J. P. (1981). Nanosecond four-wave mixing and holography in BSO crystals. *Applied Optics*, **20**, 2173–5.

Herriau, J. P., Huignard, J. P. & Aubourg, P. (1978). Some polarization properties of volume holograms in $Bi_{12}SiO_{20}$ crystals and applications. *Applied Optics*, **17**, 1851–2.

Hildebrand, B. P. & Haines, K. A. (1966). The generation of 3-dimensional contour maps by wavefront reconstruction. *Physics Letters*, **21**, 422–3.

Hildebrand, B. P. & Haines, K. A. (1967). Multiple-wavelength and multiple-source holography applied to contour generation. *Journal of the Optical Society of America*, **57**, 155–62.

Hioki, R. & Suzuki, T. (1965). Reconstruction of wavefronts in all directions. *Japanese Journal of Applied Physics*, **4**, 816.

Holloway, D. C. & Johnson, R. H. (1971). Advancements in holographic photoelasticity. *Experimental Mechanics*, **11**, 57–63.

Honda, T., Okada, K. & Tsujiuchi, J. (1981). 3-D distortion of observed images reconstructed from a cylindrical holographic stereogram: (1) laser light reconstruction type. *Optics Communications*, **36**, 11–16.

Hopf, F. A. (1980). Interferometry using conjugate-wave generation. *Journal of the Optical Society of America*, **70**, 1320–3.

Hopkins, H. H. (1950). *Wave Theory of Aberrations*. Oxford: The Clarendon Press.

Hou, S. L. & Oliver, D. S. (1971). Pockels readout optical memory using $Bi_{12}SiO_{20}$. *Applied Physics Letters*, **18**, 325–8.

Hovanesian, J. D. (1974). Variable isochromatic/isopachic fringe visibility in photo-holoelasticity. *Experimental Mechanics*, **14**, 233–6.

Hovanesian, J. D., Brcic, V. & Powell, R. L. (1968). A new stress-optic method: stress-holo interferometry. *Experimental Mechanics*, **8**, 362–8.

Huff, L. & Fusek, R. L. (1980). Color holographic stereograms. *Optical Engineering*, **19**, 691–5.

Huff, L. & Fusek, R. L. (1981). Optical techniques for increasing image width in cylindrical holographic stereograms. *Optical Engineering*, **20**, 241–5.

Huignard, J. P. (1981). Phase conjugation, real time holography and degenerate four-wave mixing in photoreactive BSO crystals. In *Current Trends in Optics*, ed. F. T. Arecchi and F. R. Aussenegg, pp. 150–60. London: Taylor & Francis.

Huignard, J. P., Herriau, J. P., Aubourg, P. & Spitz, E. (1979). Phase-conjugate wavefront generation via real-time holography in $Bi_{12}SiO_{20}$ crystals. *Optics Letters*, **4**, 21–3.

Huignard, J. P., Herriau, J. P., Pichon, L. & Marrakchi, A. (1980). Speckle-free imaging in four-wave mixing experiments with $Bi_{12}SiO_{20}$ crystals. *Optics Letters*, **5**, 436–7.

Huignard, J. P., Herriau, J. P., Rivet, G. & Günter, P. (1980). Phase conjugation and spatial frequency dependence of wavefront reflectivity in $Bi_{12}SiO_{20}$ crystals. *Optics Letters*, **5**, 102–4.

Huignard, J. P., Herriau, J. P. & Valentin, T. (1977). Time average holographic interferometry with photoconductive electrooptic $Bi_{12}SiO_{20}$ crystals. *Applied Optics*, **16**, 2796–8.

Huignard, J. P. & Micheron, F. (1976). High sensitivity read–write volume holographic storage in $Bi_{12}SiO_{20}$ and $Bi_{12}GeO_{20}$ crystals. *Applied Physics Letters*, **29**, 591–3.

Hung, Y. Y., Hu, C. P., Henley, D. R. & Taylor, C. E. (1973). Two improved methods of surface displacement measurements by holographic interferometry. *Optics Communications*, **8**, 48–51.

Hutley, M. C. (1975). Blazed interference diffraction gratings for the ultraviolet. *Optica Acta*, **22**, 1–13.

Hutley, M. C. (1976). Interference (holographic) diffraction gratings. *Journal of Physics E: Scientific Instruments*, **9**, 513–20.

Hutley, M. C. (1982). *Diffraction Gratings*. London: Academic Press.

Ichioka, Y. & Lohmann, A. W. (1972). Interference testing of large optical components with circular computer holograms. *Applied Optics*, **11**, 2597–602.

Ih, C. S. (1975). Multicolor imagery from holograms by spatial filtering. *Applied Optics*, **14**, 438–44.

Ih, C. S. (1977). Holographic laser beam scanners utilizing an auxiliary reflector. *Applied Optics*, **16**, 2137–46.

Ih, C. S. & Baxter, L. A. (1978). Improved random spatial phase modulation for speckle elimination. *Applied Optics*, **17**, 1447–54.

Ineichen, B., Liegeois, C. & Meyrueis, P. (1982). Thermoplastic film camera for holographic recording of extended objects in industrial applications. *Applied Optics*, **21**, 2209–14.

Ja, Y. H. (1982). Observation of interference between a signal and its conjugate in a four-wave mixing experiment using $Bi_{12}GeO_{20}$ crystals. *Optical & Quantum Electronics*, **14**, 367–9.

Jacobson, A. D. & McClung, F. J. (1965). Holograms produced with pulsed laser illumination. *Applied Optics*, **4**, 1509–10.

Jahoda, F. C., Jeffries, R. A. & Sawyer, G. A. (1967). Fractional fringe holographic plasma interferometry. *Applied Optics*, **6**, 1407–10.

Jannson, T. (1974). Impulse response and Shannon number of holographic optical systems. *Optics Communications*, **10**, 232–7.

Janta, J. & Miler, M. (1972). Time-average holographic interferometry of damped oscillations. *Optik*, **36**, 185–95.

Jeong, T. H. (1967). Cylindrical holography and some proposed applications. *Journal of the Optical Society of America*, **57**, 1396–8.

Jeong, T. H., Rudolf, P. & Luckett, A. (1966). 360° holography. *Journal of the Optical Society of America*, **56**, 1263–4.

Jo, J. C. & Lee, S. S. (1982). Holographic image restoration by using an unconstrained single deblurring filter. *Optica Acta*, **29**, 1231–6.

Joly, L. & Vanhorebeek, R. (1980). Development effects in white-light reflection holography. *Photographic Science & Engineering*, **24**, 108–13.

Jones, M. I., Walkup, J. F. & Hagler, M. O. (1982). Multiplex hologram representations of space-variant optical systems using ground-glass encoded reference beams. *Applied Optics*, **21**, 1291–7.

Jordan, M. P. & Solymar, L. (1978). A note on volume holograms. *Electronics Letters*, **14**, 271–2.

Kalestynski, A. (1973). Holographic multiplication in one exposure by the use of a multibeam reference field. *Applied Optics*, **12**, 1946–50.

Kalestynski, A. (1976). Multiplying lensless Fourier holograms recorded using a multibeam reference light field. *Applied Optics*, **15**, 853–5.

Kamshilin, A. A. & Miteva, M. G. (1981). Effect of infra-red irradiation on holographic recording in bismuth silicon oxide. *Optics Communications*, **36**, 429–33.

Kasahara, T., Kimura, Y., Hioki, R. & Tanaka, S. (1969). Stereoradiography using holographic techniques. *Japanese Journal of Applied Physics*, **8**, 124–5.

Kaspar, F. G. (1973). Diffraction by thick periodically stratified gratings with complex dielectric constant. *Journal of the Optical Society of America*, **63**, 37–45.

Kaspar, F. G. (1974). Computation of light transmitted by a thick grating for application to contact printing. *Journal of the Optical Society of America*, **64**, 1623–30.

Kato, M. & Okino, Y. (1973). Speckle reduction by double recorded holograms. *Applied Optics*, **12**, 1199–201.

Kermisch, D. (1969). Nonuniform sinusoidally modulated dielectric gratings. *Journal of the Optical Society of America*, **59**, 1409–14.

Kermisch, D. (1970). Image reconstruction from phase information only. *Journal of the Optical Society of America*, **60**, 15–17.

Kermisch, D. (1971). Efficiency of photochromic gratings. *Journal of the Optical Society of America*, **61**, 1202–6.

Kessler, S. & Kowarschik, R. (1975). Diffraction efficiency of volume holograms. Part 1. Transmission holograms. *Optical & Quantum Electronics*, **7**, 1–14.

Kiemle, U. (1974). Considerations on holographic memories in the gigabyte region. *Applied Optics*, **13**, 803–7.

Killat, U. (1977a). Coupled wave theory of hologram gratings with arbitrary attenuation. *Optics Communications*, **21**, 110–11.

Killat, U. (1977b). Holographic microfiche of picture-like information. *Optica Acta*, **24**, 453–62.

King, M. C. (1968). Multiple exposure hologram recording of a 3-D image with a 360° view. *Applied Optics*, **7**, 1641–2.

King, M. C., Noll, A. M. & Berry, D. H. (1970). A new approach to computer generated holography. *Applied Optics*, **9**, 471–5.

Klein, W. R. & Cook, B. D. (1967). Unified approach to ultrasonic light diffraction. *IEEE Transactions on Sonics & Ultrasonics*, SU-14, 123–34.

Klimenko, I. S., Matinyan, E. G. & Dubitskii, L. G. (1975). Use of focused image holography for the nondestructive testing of electronic parts. *Soviet Journal of Nondestructive Testing*, **10**, 696–9.

Knight, G. (1974). Page-oriented associative holographic memory. *Applied Optics*, **13**, 904–12.

Knight, G. (1975*a*). Holographic associative memory and processor. *Applied Optics*, **14**, 1088–92.

Knight, G. R. (1975*b*). Holographic memories. *Optical Engineering*, **14**, 453–9.

Kock, M. & Tiemens, U. (1973). Tomosynthesis: a holographic method for variable depth display. *Optics Communications*, **7**, 260–5.

Koechner, W. (1979*a*). Solid state lasers. In *Handbook of Optical Holography*, ed. H. J. Caulfield, pp. 257–67. New York: Academic Press.

Koechner, W. (1979*b*). Pulsed holography. In *Laser Handbook*, ed. M. L. Stitch, pp. 578–626. Amsterdam: North-Holland.

Kogelnik, H. (1965). Holographic image projection through inhomogeneous media. *Bell System Technical Journal*, **44**, 2451–5.

Kogelnik, H. (1967). Reconstructing response and efficiency of hologram gratings. In *Proceedings of the Symposium on Modern Optics*, pp. 605–17. Brooklyn: Polytechnic Press.

Kogelnik, H. (1969). Coupled wave theory for thick hologram gratings. *Bell System Technical Journal*, **48**, 2909–47.

Kogelnik, H. (1972). Optics at Bell Laboratories – lasers in technology. *Applied Optics*, **11**, 2426–34.

Komar, V. G. (1977). Progress on the holographic movie process in the USSR. In *Three-Dimensional Imaging*, Proceedings of the SPIE, vol. 120, ed. S. A. Benton, pp. 127–44. Redondo Beach: SPIE.

Kowarschik, R. (1976). Diffraction efficiency of attenuated sinusoidally modulated gratings in volume holograms. *Optica Acta*, **23**, 1039–51.

Kowarschik, R. (1978*a*). Diffraction efficiency of sequentially stored gratings in transmission volume holograms. *Optica Acta*, **25**, 67–81.

Kowarschik, R. (1978*b*). Diffraction efficiency of sequentially stored gratings in reflection volume holograms. *Optical & Quantum Electronics*, **10**, 171–8.

Kowarschik, R. & Kessler, S. (1975). Zum Beugungswirkungsgrad von Volumenholog-rammen. Teil II. Reflexionshologramme. *Optical & Quantum Electronics*, **7**, 399–411.

Kozma, A. (1966). Photographic recording of spatially modulated coherent light. *Journal of the Optical Society of America*, **56**, 428–32.

Kozma, A. (1968*a*). Effects of film grain noise in holography. *Journal of the Optical Society of America*, **58**, 436–8.

Kozma, A. (1968*b*). Analysis of the film non-linearities in hologram recording. *Optica Acta*, **15**, 527–51.

Kozma, A., Jull, G. W. & Hill, K. O. (1970). An analytical and experimental study of non-linearities in hologram recording. *Applied Optics*, **9**, 721–31.

Kozma, A. & Massey, N. (1969). Bias level reduction of incoherent holograms. *Applied Optics*, **8**, 393–7.

Kozma, A. & Zelenka, J. S. (1970). Effect of film resolution and size in holography. *Journal of the Optical Society of America*, **60**, 34–43.

Kramer, C. J. (1981). Holographic laser scanners for nonimpact printing. *Laser Focus*, **17**, No. 6, 70–82.

Kreis, T. M. & Kreitlow, H. (1980). Digital processing of holographic interference patterns. In *Technical Digest on Hologram Interferometry & Speckle Metrology*, pp. TuB2-1–TuB2-4. Washington: The Optical Society of America.

Krile, T. F., Marks II, R. J., Walkup, J. F. & Hagler, M. O. (1977). Holographic representations of space-variant systems using phase-coded reference beams. *Applied Optics*, **16**, 3131–5.

Kubo, H. & Nagata, R. (1976*a*). Holographic photoelasticity with depolarized object wave. *Japanese Journal of Applied Physics*, **15**, 641–4.

Kubo, H. & Nagata, R. (1976*b*). Application of polarization holography by the Kurtz's method to photoelasticity. *Japanese Journal of Applied Physics*, **15**, 1095–9.

Kubota, K., Ono, Y., Kondo, M., Sugama, S., Nishida, N. & Sakaguchi, M. (1980). Holographic disk with high data transfer rate: its application to an audio response memory. *Applied Optics*, **19**, 944–51.

Kubota, T. (1978). Characteristics of thick hologram gratings recorded in absorptive medium. *Optica Acta*, **25**, 1035–53.

Kubota, T. & Ose, T. (1979*a*). Methods of increasing the sensitivity of methylene blue sensitized dichromated gelatin. *Applied Optics*, **18**, 2538–9.

Kubota, T. & Ose, T. (1979*b*). Lippmann color holograms recorded in methylene-blue sensitized dichromated gelatin. *Optics Letters*, **4**, 289–91.

Küchel, F. M. & Tiziani, H. J. (1981). Real-time contour holography using BSO crystals. *Optics Communications*, **38**, 17–20.

Kurtz, C. N. (1968). Copying reflection holograms. *Journal of the Optical Society of America*, **58**, 856–7.

Kurtz, C. N. (1969). Holographic polarization recording with an encoded reference beam. *Applied Physics Letters*, **14**, 59–61.

Labeyrie, A. & Flamand, J. (1969). Spectrographic performance of holographically made diffraction gratings. *Optics Communications*, **1**, 5–8.

La Macchia, J. T. & White, D. L. (1968). Coded multiple-exposure holograms. *Applied Optics*, **7**, 91–4.

Lamberts, R. L. & Kurtz, C. N. (1971). Reversal bleaching for low flare light in holograms. *Applied Optics*, **19**, 1342–7.

Landry, M. J. (1967). The effect of two hologram-copying parameters on the quality of copies. *Applied Optics*, **6**, 1947–56.

Lang, M. & Eschler, H. (1974). Gigabyte capacities for holographic memories. *Optics & Laser Technology*, **6**, 219–24.

Langbein, U. & Lederer, F. (1980). Modal theory for thick holographic gratings with sharp boundaries. I. General treatment. *Optica Acta*, **27**, 171–82.

Langdon, R. M. (1970). A high capacity holographic memory. *The Marconi Review*, **33**, 113–30.

Latta, J. N. (1971*a*). Computer-based analysis of hologram imagery and aberrations. I. Hologram types and their nonchromatic aberrations. *Applied Optics*, **10**, 599–608.

Latta, J. N. (1971*b*). Computer-based analysis of hologram imagery and aberrations. II. Aberrations induced by a wavelength shift. *Applied Optics*, **10**, 609–18.

Latta, J. N. (1971*c*). Computer-based analysis of holography using ray tracing. *Applied Optics*, **10**, 2698–710.

Latta, M. R. & Pole, R. V. (1979). Design techniques for forming 488 nm holographic lenses with reconstruction at 633 nm. *Applied Optics*, **18**, 2418–21.

Lederer, F. & Langbein, U. (1977). Attenuated thick hologram gratings. Part I. Diffraction efficiency. *Optical & Quantum Electronics*, **9**, 473–85.

Lederer, F. & Langbein, U. (1980). Modal theory for thick holographic gratings with sharp boundaries. II. Unslanted transmission and reflection gratings. *Optica Acta*, **27**, 183–200.

Lee, S. H. ed. (1981). *Optical Information Processing: Fundamentals*, Topics in Applied Physics, vol. 48. Berlin: Springer-Verlag.

Lee, T. C., Lin, J. W. & Tufte, O. N. (1977). Thermoplastic photoconductor for optical recording and storage – new developments. In *Optical Storage Materials & Methods*, Proceedings of the SPIE, vol. 123, ed. L. Beiser & D. Chen, pp. 74–7. Bellingham: SPIE.

Lee, W. H. (1970). Sampled Fourier transform hologram generated by computer. *Applied Optics*, **9**, 639–43.

Lee, W. H. (1974). Binary synthetic holograms. *Applied Optics*, **13**, 1677–82.

Lee, W. H. (1978). Computer-generated holograms: Techniques and applications. In *Progress in Optics*, vol. 16, ed. E. Wolf, pp. 121–232. Amsterdam: North-Holland.

Lee, W. H. (1979). Binary computer-generated holograms. *Applied Optics*, **18**, 3661–9.

Lee, W. H. & Streifer, W. (1978a). Diffraction efficiency of evanescent-wave holograms. I. TE polarization. *Journal of the Optical Society of America*, **68**, 795–801.

Lee, W. H. & Streifer, W. (1978b). Diffraction efficiency of evanescent-wave holograms. II. TM polarization. *Journal of the Optical Society of America*, **68**, 802–9.

Leith, E. N. & Chang, B. J. (1973). Space invariant holography with quasi-coherent light. *Applied Optics*, **12**, 1957–63.

Leith, E. N. & Chen, H. (1978). Deep image rainbow holograms. *Optics Letters*, **2**, 82–4.

Leith, E. N., Chen, H. & Roth, J. (1978). White light hologram technique. *Applied Optics*, **17**, 3187–8.

Leith, E. N., Kozma, A., Upatnieks, J., Marks, J. & Massey, N. (1966). Holographic data storage in three-dimensional media. *Applied Optics*, **5**, 1303–12.

Leith, E. N. & Upatnieks, J. (1962). Reconstructed wavefronts and communication theory. *Journal of the Optical Society of America*, **52**, 1123–30.

Leith, E. N. & Upatnieks, J. (1963). Wavefront reconstruction with continuous-tone objects. *Journal of the Optical Society of America*, **53**, 1377–81.

Leith, E. N. & Upatnieks, J. (1964). Wavefront reconstruction with diffused illumination and three-dimensional objects. *Journal of the Optical Society of America*, **54**, 1295–301.

Leith, E. N. & Upatnieks, J. (1965). Microscopy by wavefront reconstruction. *Journal of the Optical Society of America*, **55**, 569–70.

Leith, E. N. & Upatnieks, J. (1966). Holographic imagery through diffusing media. *Journal of the Optical Society of America*, **56**, 523.

Leith, E. N. & Upatnieks, J. (1967). Holography with achromatic fringe systems. *Journal of the Optical Society of America*, **57**, 975–80.

Leith, E. N., Upatnieks, J. & Haines, K. A. (1965). Microscopy by wavefront reconstruction. *Journal of the Optical Society of America*, **55**, 981–6.

Lengyel, B. A. (1971). *Lasers*. New York: Wiley-Interscience.

Lesem, L. B., Hirsch, P. M. & Jordan Jr, J. A. (1969). The kinoform: a new wavefront reconstruction device. *IBM Journal of Research & Development*, **13**, 150–5.

Lessard, R. A., Som, S. C. & Boivin, A. (1973). New technique of color holography. *Applied Optics*, **12**, 2009–11.

Leung, K. M., Lee, T. C., Bernal, E. & Wyant, J. C. (1979). Two-wavelength contouring with the automated thermoplastic holographic camera. In *Interferometry*, Proceedings of the SPIE, vol. 192, ed. G. W. Hopkins, pp. 184–9. Bellingham: SPIE.

Leung, K. M., Lindquist, J. C. & Shepherd, L. T. (1980). E-beam computer generated holograms for aspheric testing. In *Recent Advances in Holography*, Proceedings of the SPIE, vol. 215, ed. T. C. Lee & P. N. Tamura, pp. 70–5. Bellingham: SPIE.

Levitt, J. A. & Stetson, K. A. (1976). Mechanical vibrations: mapping their phase with hologram interferometry. *Applied Optics*, **15**, 195–9.

Lin, L. H. (1969). Hologram formation in hardened dichromated gelatin films. *Applied Optics*, **8**, 963–6.

Lin, L. H. (1971). Method of characterizing hologram-recording materials. *Journal of the Optical Society of America*, **61**, 203–8.

Lin, L. H. & Beauchamp, H. L. (1970a). An automatic shutter for holography. *Review of Scientific Instruments*, **41**, 1438–40.

Lin, L. H. & Beauchamp, H. L. (1970b). Write–read–erase in situ optical memory using thermoplastic holograms. *Applied Optics*, **9**, 2088–92.

Lin, L. H. & Lo Bianco, C. V. (1967). Experimental techniques in making multicolor white light reconstructed holograms. *Applied Optics*, **6**, 1255–8.

Lin, L. H., Pennington, K. S., Stroke, G. W. & Labeyrie, A. E. (1966). Multicolor holographic image reconstruction with white light illumination. *Bell System Technical Journal*, **45**, 659–60.

Livanos, A. C., Katzir, A., Shellan, J. B. & Yariv, A. (1977). Linearity and enhanced sensitivity of the Shipley AZ-1350 B photoresist. *Applied Optics*, **16**, 1633–5.

Loewen, E., Maystre, D., McPhedran, R. & Wilson, I. (1975). Correlation between efficiency of diffraction gratings and theoretical calculations over a wide range. *Japanese Journal of Applied Physics*, **14** (Supplement 14-1), 143–52.

Lohmann, A. (1956). Optische Einseitenbandübertragung angewandt auf das Gabor-Mikroskop. *Optica Acta*, **3**, 97–9.

Lohmann, A. W. (1965a). Reconstruction of vectorial wavefronts. *Applied Optics*, **4**, 1667–8.

Lohmann, A. W. (1965b). Wavefront reconstruction for incoherent objects. *Journal of the Optical Society of America*, **55**, 1555–6.

Lohmann, A. W. & Paris, D. P. (1967). Binary Fraunhofer holograms generated by computer. *Applied Optics*, **6**, 1739–48.

Loomis, J. S. (1980). Computer-generated holography and optical testing. *Optical Engineering*, **19**, 679–85.

Lowenthal, S. & Joyeux, D. (1971). Speckle removal by a slowly moving diffuser associated with a motionless diffuser. *Journal of the Optical Society of America*, **61**, 847–51.

Lowenthal, S., Serres, J. & Froehly, C. (1969). Enregistrement d'hologrammes en lumière spatialement incohérente. *Comptes Rendus des Seances de l'Academie des Sciences, Paris, B*, **268**, 841–4.

Lu, S. (1968). Generating multiple images for integrated circuits by Fourier-transform holograms. *Proceedings of the IEEE*, **56**, 116–17.

Lukin, A. V. & Mustafin, K. S. (1979). Holographic methods of testing aspherical surfaces. *Soviet Journal of Optical Technology*, **46**, 237–44.

Lukosz, W. & Wüthrich, A. (1974). Holography with evanescent waves. I. Theory of the diffraction efficiency for s-polarized light. *Optik*, **41**, 191–211.

Lukosz, W. & Wüthrich, A. (1976). Hologram recording and read-out with the evanescent field of guided waves. *Optics Communications*, **19**, 232–5.

MacGovern, A. J. & Wyant, J. C. (1971). Computer-generated holograms for testing optical elements. *Applied Optics*, **10**, 619–24.

MacQuigg, D. R. (1977). Hologram fringe stabilization method. *Applied Optics*, **16**, 291–2.

Magnusson, R. & Gaylord, T. K. (1977). Analysis of multiwave diffraction of thick gratings. *Journal of the Optical Society of America*, **67**, 1165–70.

Magnusson, R. & Gaylord, T. K. (1978a). Diffraction regimes of transmission gratings. *Journal of the Optical Society of America*, **68**, 809–14.

Magnusson, R. & Gaylord, T. K. (1978b). Equivalence of multiwave coupled-wave theory and modal theory for periodic-media diffraction. *Journal of the Optical Society of America*, **68**, 1777–9.

Malin, M. & Morrow, H. E. (1981). Wavelength scaling holographic elements. *Optical Engineering*, **20**, 756–8.

Mallick, S. (1975). Pulse holography of uniformly moving objects. *Applied Optics*, **14**, 602–5.

Mandel, L. (1965). Color imagery by wavefront reconstruction. *Journal of the Optical Society of America*, **55**, 1697–8.

Maréchal, A. & Croce, P. (1953). A filter of spatial frequencies for the improvement of contrast of optical images. *Comptes Rendus des Seances de l'Academie des Sciences*, Serie II, **237**, 607–9.

Marom, E. (1967). Color imagery by wavefront reconstruction. *Journal of the Optical Society of America*, **57**, 101–2.

Marrakchi, A., Huignard, J. P. & Herriau, J. P. (1980). Application of phase conjugation in $Bi_{12}SiO_{20}$ crystals to mode pattern visualisation of diffuse vibrating structures. *Optics Communications*, **34**, 15–18.

Marrone, E. S. & Ribbens, W. B. (1975). Dual-index holographic contour mapping over a large range of contour spacings. *Applied Optics*, **14**, 23–4.

Martienssen, W. & Spiller, S. (1967). Holographic reconstruction without granulation. *Physics Letters*, **24A**, 126–8.

Matsuda, K., Freund, C. H. & Hariharan, P. (1981). Phase difference amplification using longitudinally reversed shearing interferometry: an experimental study. *Applied Optics*, **20**, 2763–5.

Matsumoto, T., Iwata, K. & Nagata, R. (1973). Measuring accuracy of three-dimensional displacements in holographic interferometry. *Applied Optics*, **12**, 961–7.

Matsumoto, K. & Takashima, M. (1970). Phase-difference amplification by nonlinear holograms. *Journal of the Optical Society of America*, **60**, 30–3.

Matsumura, M. (1975). Speckle noise reduction by random phase shifters. *Applied Optics*, **14**, 660–5.

Mazakova, M., Pancheva, M., Kandilarov, P. & Sharlandjiev, P. (1982a). Dichromated gelatin for volume holographic recording with high sensitivity. Part I. *Optical & Quantum Electronics*, **14**, 311–15.

Mazakova, M., Pancheva, M., Kandilarov, P. & Sharlandjiev, P. (1982b). Dichromated gelatin for volume holographic recording with high sensitivity. Part II. *Optical & Quantum Electronics*, **14**, 317–20.

McCauley, D. G., Simpson, C. E. & Murbach, W. J. (1973). Holographic optical element for visual display applications. *Applied Optics*, **12**, 232–42.

McClung, F. J., Jacobson, A. D. & Close, D. H. (1970). Some experiments performed with a reflected-light pulsed laser holography system. *Applied Optics*, **9**, 103–6.

McCrickerd, J. T. & George, N. (1968). Holographic stereogram from sequential component photographs. *Applied Physics Letters*, **12**, 10–12.

McKechnie, T. S. (1975a). Reduction of speckle in an image by a moving aperture: second order statistics. *Optics Communications*, **13**, 29–34.

McKechnie, T. S. (1975*b*). Reduction of speckle by a moving aperture: first order statistics. *Optics Communications*, **13**, 35–9.

McKechnie, T. S. (1975*c*). Speckle reduction. In *Laser Speckle & Related Phenomena*, Topics in Applied Physics, vol. 9, ed. J. C. Dainty, pp. 123–70. Berlin: Springer-Verlag.

McMahon, D. H., Franklin, A. R. & Thaxter, J. B. (1969). Light beam deflection using holographic scanning techniques. *Applied Optics*, **8**, 399–402.

McMahon, D. H. & Maloney, W. T. (1970). Measurements of the stability of bleached photographic phase holograms. *Applied Optics*, **9**, 1363–8.

McPhedran, R. C., Wilson, I. J. & Waterworth, M. D. (1973). Profile formation in holographic diffraction gratings. *Optics & Laser Technology*, **5**, 166–71.

Meier, R. W. (1965). Magnification and third-order aberrations in holography. *Journal of the Optical Society of America*, **55**, 987–92.

Menzel, E. (1974). Comment to the methods of contour holography. *Optik*, **40**, 557–9.

Mercier, R. & Lowenthal, S. (1980). Comparison of in-line and carrier frequency holograms in aspheric testing. *Optics Communications*, **33**, 251–6.

Mertz, L. & Young, N. O. (1962). Fresnel transformations of images. In *Proceedings of the Conference on Optical Instruments and Techniques*, London, 1961, ed. K. J. Habell, pp. 305–12. London: Chapman & Hall.

Meyerhofer, D. (1972). Phase holograms in dichromated gelatin. *RCA Review*, **33**, 110–30.

Miles, J. F. (1972). Imaging and magnification properties in holography. *Optica Acta*, **19**, 165–86.

Miles, J. F. (1973). Evaluation of the wavefront aberration in holography. *Optica Acta*, **20**, 19–31.

Miridonov, S. V., Petrov, M. P. & Stepanov, S. I. (1978). Light diffraction by volume holograms in optically active photorefractive crystals. *Soviet Technical Physics Letters*, **4**, 393–4.

Moharam. M. G. & Gaylord, T. K. (1981). Rigorous coupled-wave analysis of planar-grating diffraction, *Journal of the Optical Society of America*, **71**, 811–18.

Moharam, M. G., Gaylord, T. K. & Magnusson, R. (1980*a*). Criteria for Bragg regime diffraction by phase gratings. *Optics Communications*, **32**, 14–18.

Moharam, M. G., Gaylord, T. K. & Magnusson, R. (1980*b*). Criteria for Raman–Nath regime diffraction by phase gratings. *Optics Communications*, **32**, 19–23.

Moharam, M. G. & Young, L. (1978). Criterion for Bragg and Raman–Nath diffraction regimes. *Applied Optics*, **17**, 1757–9.

Molin, N. E. & Stetson, K. A. (1969). Measuring combination mode vibration patterns by hologram interferometry. *Journal of Physics E: Scientific Instruments*, **2**, 609–12.

Molin, N. E. & Stetson, K. A. (1970*a*). Measurement of fringe loci and localization in hologram interferometry for pivot motion, in-plane rotation and in-plane translation. Part I. *Optik*, **31**, 157–77.

Molin, N. E. & Stetson, K. A. (1970*b*). Measurement of fringe loci and localization in hologram interferometry for pivot motion, in-plane rotation and in-plane translation. Part II. *Optik*, **31**, 281–91.

Molin, N. E. & Stetson, K. A. (1971). Fringe localization in hologram interferometry of mutually independent and dependent rotations around orthogonal, non-intersecting axes. *Optik*, **33**, 399–422.

Mosyakin, Yu. S. & Skrotskii, G. V. (1972). Holographic optical elements. *Soviet Journal of Quantum Electronics*, **2**, 199–206.

Mottier, F. M. (1969). Holography of randomly moving objects. *Applied Physics Letters*, **15**, 44–5.

Nakadate, S., Magome, N., Honda, T. & Tsujiuchi, J. (1981). Hybrid holographic interferometer for measuring three-dimensional deformations. *Optical Engineering*, **20**, 246–52.

Namioka, T., Seya, M. & Noda, H. (1976). Design and performance of holographic concave gratings. *Japanese Journal of Applied Physics*, **15**, 1181–97.

Nassenstein, H. (1968*a*). Copying of holograms. *Optik*, **27**, 327–34.

Nassenstein, H. (1968*b*). Holographie und Interferenzversuche mit inhomogenen Oberflächenwellen. *Physics Letters*, **28A**, 249–51.

Nassenstein, H. (1969). Rekonstruktion von Hologrammen mit höherem Beugungs-wirkungsgrad. *Optik*, **30**, 201–5.

Nassenstein, H. (1970). Superresolution by diffraction of subwaves. *Optics Communications*, **2**, 231–4.

Nath, N. S. N. (1938). Diffraction of light by supersonic waves. *Proceedings of the Indian Academy of Sciences*, **8A**, 499–503.

Nelson, R. H., Vander Lugt, A. & Zech, R. G. (1974). Holographic data storage and retrieval. *Optical Engineering*, **13**, 429–30.

Neumann, D. B. (1966). Geometrical relationships between the original object and the two images of a hologram reconstruction. *Journal of the Optical Society of America*, **56**, 858–61.

Neumann, D. B. (1968). Holography of moving scenes. *Journal of the Optical Society of America*, **58**, 447–54.

Neumann, D. B., Jacobson, C. F. & Brown, G. M. (1970). Holographic technique for determining the phase of vibrating objects. *Applied Optics*, **9**, 1357–68.

Neumann, D. B. & Penn, R. C. (1972). Object motion compensation using reflection holography. *Journal of the Optical Society of America*, **62**, 1373.

Neumann, D. B. & Rose, H. W. (1967). Improvement of recorded holographic fringes by feedback control. *Applied Optics*, **6**, 1097–1104.

Nishihara, H. & Koyama, J. (1979). New technique to record and playback holographic color video memories. *Optics Communications*, **31**, 16–20.

Nisida, M. & Saito, H. (1964). A new interferometric method of two-dimensional stress analysis. *Experimental Mechanics*, **4**, 366–76.

Nobis, D. & Vest, C. M. (1978). Statistical analysis of errors in holographic interferometry. *Applied Optics*, **17**, 2198–204.

Norman, S. L. & Singh, M. P. (1975). Spectral sensitivity and linearity of Shipley AZ-1350J photoresist. *Applied Optics*, **14**, 818–20.

Okada, K., Honda, T. & Tsujiuchi, J. (1981). 3-d distortion of observed images reconstructed from a cylindrical holographic stereogram (2) white light reconstruction type. *Optics Communications*, **36**, 17–21.

Okada, K., Honda, T. & Tsujiuchi, J. (1982). Image blur of multiplex holograms. *Optics Communications*, **41**, 397–402.

Okoshi, T. (1977). Projection-type holography. In *Progress in Optics*, vol. 15, ed. E. Wolf, pp. 141–85. Amsterdam: North-Holland.

O'Neill, E. L. (1956). Spatial filtering in optics. *IEEE Transactions on Information Theory*, **IT-2**, 56–65.

O'Neill, E. L. (1963). *Introduction to Statistical Optics*. Reading: Addison-Wesley Publishing Company.

Oreb, B. F. & Hariharan, P. (1981). Improved integrating exposure-control system for color holography. *Optical Engineering*, **20**, 749–52.

O'Regan, R. & Dudderar, T. D. (1971). A new holographic interferometer for stress analysis. *Experimental Mechanics*, **11**, 241–7.

Östlund, L. A. & Biedermann, K. (1977). Laser speckle reduction: equivalence of the moving aperture method and incoherent spatial filtering. *Applied Optics*, **16**, 685–90.

Ostrovskaya, G. V. & Ostrovskii, Yu. I. (1971). Two-wavelength hologram method for studying the dispersion properties of phase objects. *Soviet Physics – Technical Physics*, **15**, 1890–2.

Owen, M. P. & Solymar, L. (1980). Efficiency of volume phase reflection holograms recorded in an attenuating medium. *Optics Communications*, **34**, 321–6.

Palais, J. C. & Wise, J. A. (1971). Improving the efficiency of very low efficiency holograms by copying. *Applied Optics*, **10**, 667–8.

Papoulis, A. (1962). *The Fourier Integral & its Applications*. New York: McGraw-Hill.

Papoulis, A. (1965). *Probability, Random Variables & Stochastic Processes*. New York: McGraw Hill.

Parker, R. J. (1978). A new method of frozen-fringe holographic interferometry using thermoplastic recording media. *Optica Acta*, **25**, 787–92.

Pastor, J. (1969). Hologram interferometry and optical technology. *Applied Optics*, **8**, 525–31.

Pennington, K. S., Harper, J. S. & Laming, F. P. (1971). New phototechnology suitable for recording phase holograms and similar information in hardened gelatine. *Applied Physics Letters*, **18**, 80–4.

Pennington, K. S. & Lin, L. H. (1965). Multicolor wavefront reconstruction. *Applied Physics Letters*, **7**, 56–7.

Petrov, M. P., Stepanov, S. I. & Kamshilin, A. A. (1979*a*). Light diffraction from volume holograms in electro-optic birefringent crystals. *Optics Communications*, **29**, 44–8.

Petrov, M. P., Stepanov, S. I. & Kamshilin, A. A. (1979*b*). Holographic storage of information and peculiarities of light diffraction in birefringent electro-optic crystals. *Optical & Laser Technology*, **11**, 149–51.

Phillips, N. J., Ward, A. A., Cullen, R. & Porter, D. (1980). Advances in holographic bleaches. *Photographic Science & Engineering*, **24**, 120–4.

Pole, R. V., Werlick, H. W. & Krusche, R. J. (1978). Holographic light deflection. *Applied Optics*, **17**, 3294–7.

Pole, R. V. & Wolenmann, H. P. (1975). Holographic laser beam deflector. *Applied Optics*, **14**, 976–80.

Powell, R. L. & Stetson, K. A. (1965). Interferometric vibration analysis by wavefront reconstruction. *Journal of the Optical Society of America*, **55**, 1593-8.

Prikryl, I. & Kvapil, J. (1980). A note on hologram synthesis from 2-D transparencies. *Journal of Optics (Paris)*, **11**, 231–3.

Pryputniewicz, R. J. (1978). Holographic strain analysis: an experimental implementation of the fringe vector theory. *Applied Optics*, **17**, 3613–18.

Pryputniewicz, R. J. (1980). The properties of fringes in hologram interferometry. In *Technical Digest on Hologram Interferometry & Speckle Metrology*, pp. MB1-1–MB1-7. Washington: The Optical Society of America.

Pryputniewicz, R. J. & Bowley, W. W. (1978). Techniques of holographic displacement measurement: an experimental comparison. *Applied Optics*, **17**, 1748–56.

Pryputniewicz, R. J. & Stetson, K. A. (1976). Holographic strain analysis: extension of fringe-vector method to include perspective. *Applied Optics*, **15**, 725–8.

Pryputniewicz, R. J. & Stetson, K. A. (1980). Determination of sensitivity vectors in hologram interferometry from two known rotations of the object. *Applied Optics*, **19**, 2201–5.

Radley Jr, R. J. (1975). Two-wavelength holography for measuring plasma electron density. *Physics of Fluids*, **18**, 175–9.

Ragnarsson, S. I. (1970). A new holographic method of generating a high efficiency, extended range spatial filter with application to restoration of defocussed images. *Physica Scripta*, **2**, 145–53.

Ragnarsson, S. I. (1978). Scattering phenomena in volume holograms with strong coupling. *Applied Optics*, **17**, 116–27.

Rau, J. E. (1966). Detection of differences in real distributions. *Journal of the Optical Society of America*, **56**, 1490–4.

Redman, J. D., Wolton, W. P. & Shuttleworth, E. (1968). Use of holography to make truly three-dimensional x-ray images. *Nature*, **220**, 58–60.

Reich, S., Rav-Noy, Z. & Friesem, A. A. (1977). Frost suppression in photoconductor-thermoplastic holographic recording devices. *Applied Physics Letters*, **31**, 654–6.

Rhodes, W. T. (1981). Space-variant optical systems and processing. In *Applications of the Optical Fourier Transform*, ed. F. Stark, pp. 333–69. New York: Academic Press.

Roberts, H. N., Watkins, J. W. & Johnson, R. H. (1974). High-speed holographic digital recorder. *Applied Optics*, **13**, 841–56.

Rogers, G. L. (1952). Experiments in diffraction microscopy. *Proceedings of the Royal Society of Edinburgh*, **63A**, 193–221.

Rogers, G. L. (1966). Polarization effects in holography. *Journal of the Optical Society of America*, **56**, 831.

Rosen, L. (1966). Focused-image holography with extended sources. *Applied Physics Letters*, **9**, 337–9.

Rosen, L. (1967). The pseudoscopic inversion of holograms. *Proceedings of the IEEE*, **55**, 118.

Rotz, F. B. & Friesem, A. A. (1966). Holograms with non-pseudoscopic real images. *Applied Physics Letters*, **8**, 146–8.

Rowley, D. M. (1979). A holographic interference camera. *Journal of Physics E: Scientific Instruments*, **12**, 971–5.

Rowley, D. M. (1981). Interferometry with miniature format volume reflection holograms. *Optica Acta*, **28**, 907–15.

Rudolph, D. & Schmahl, G. (1967). Verfahren zur Herstellung von Röntgenlinsen und Beugungsgittern. *Umschau in Wissenschaft und Technik*, **7**, 225.

Russell, P. St. J. (1981). Optical volume holography. *Physics Reports*, **71**, 209–312.

Russell, P. St. J. & Solymar, L. (1980). Borrmann-like anomalous effects in volume holography. *Applied Physics*, **22**, 335–53.

Růžek, J. & Fiala, P. (1979). Reflection holographic portraits. *Optica Acta*, **26**, 1257–64.

Saito, T., Imamura, T., Honda, T. & Tsujiuchi, J. (1980). Solvent vapour method in thermoplastic photoconductor media. *Journal of Optics* (*Paris*), **11**, 285–92.

Saito, T., Imamura, T., Honda, T. & Tsujiuchi, J. (1981). Enhancement of sensitivity by stratifying a photoconductor on thermoplastic-photoconductor media. *Journal of Optics* (*Paris*), **12**, 49–58.

Salminen, O. & Keinonen, T. (1982). On absorption and refractive index modulations of dichromated gelatin gratings. *Optica Acta*, **29**, 531–40.

Sanford, R. J. & Durelli, A. J. (1971). Interpretation of fringes in stress-holo-interferometry. *Experimental Mechanics*, **11**, 161–6.

Sato, T., Ogawa, H. & Ueda, M. (1974). Contour generation of vibrating object by weighted subtraction of holograms. *Applied Optics*, **13**, 1280–2.

Schlüter, M. (1980). Analysis of holographic interferograms with a TV picture system. *Optics & Laser Technology*, **12**, 93–5.

Schmahl, G. (1975). Holographically made diffraction gratings for the visible, UV and soft X-ray region. *Journal of the Spectroscopical Society of Japan*, **23**, Supplement No. 1, 3–11.

Schmahl, G. & Rudolph, D. (1976). Holographic diffraction gratings. In *Progress in Optics*, vol. 14, ed. E. Wolf, pp. 196–244. Amsterdam: North-Holland.

Schwar, M. R. J., Pandya, T. P. & Weinberg, F. J. (1967). Point holograms as optical elements. *Nature*, **215**, 239–41.

Shajenko, P. & Johnson, C. D. (1968). Stroboscopic holographic interferometry. *Applied Physics Letters*, **13**, 44–6.

Shankoff, T. A. (1968). Phase holograms in dichromated gelatin. *Applied Optics*, **7**, 2101–5.

Sheridon, N. K. (1968). Production of blazed holograms. *Applied Physics Letters*, **12**, 316–18.

Sherman, G. C. (1967). Hologram copying by Gabor holography of transparencies. *Applied Optics*, **6**, 1749–53.

Shibayama, K. & Uchiyama, H. (1971). Measurement of three-dimensional displacements by hologram interferometry. *Applied Optics*, **10**, 2150–4.

Siebert, L. D. (1967). Front lighted pulse laser holography. *Applied Physics Letters*, **11**, 326–8.

Siebert, L. D. (1968). Large scene front-lighted hologram of a human subject. *Proceedings of the IEEE*, **56**, 1242–3.

Sirohi, R. S., Blume, H. & Rosenbruch, K. J. (1976). Optical testing using synthetic holograms. *Optica Acta*, **23**, 229–36.

Sjölinder, S. (1981). Dichromated gelatin and the mechanism of hologram formation. *Photographic Science & Engineering*, **25**, 112–18.

Smith, H. M. (1968). Photographic relief images. *Journal of the Optical Society of America*, **58**, 533–9.

Smith, H. M. ed. (1977). *Holographic Recording Materials*, Topics in Applied Physics, vol. 20. Berlin: Springer-Verlag.

Smith, R. W. (1977). Astigmatism-free holographic lens elements. *Optics Communications*, **21**, 102–5.

Snow, K. & Vandewarker, R. (1968). An application of holography to interference microscopy. *Applied Optics*, **7**, 549–54.

Snow, K. & Vandewarker, R. (1970). On using holograms for test glasses. *Applied Optics*, **9**, 822–7.

Sollid, J. E. (1969). Holographic interferometry applied to measurements of small static displacements of diffusely reflecting surfaces. *Applied Optics*, **8**, 1587–95.

Solymar, L. (1977). A general two-dimensional theory for volume holograms. *Applied Physics Letters*, **31**, 820–2.

Solymar, L. (1978). A two-dimensional volume hologram theory including the effect of varying average dielectric constant. *Optics Communications*, **26**, 158–60.

Solymar, L. & Cooke, D. J. (1981). *Volume Holography & Volume Gratings*. New York: Academic Press.

Som, S. C. & Budhiraja, C. J. (1975). Noise reduction by continuous addition of subchannel holograms. *Applied Optics*, **14**, 1702–5.

Sopori, B. L. & Chang, W. S. C. (1971). 3-D hologram synthesis from 2-D pictures. *Applied Optics*, **10**, 2789–90.

Spitz, E. (1967). Holographic reconstruction of objects through a diffusing medium in motion. *Comptes Rendus de l'Academie des Sciences B*, **264**, 1449–51.

Staebler, D. L. (1977). Ferroelectric crystals. In *Holographic Recording Materials*, Topics in Applied Physics, vol. 20, ed. H. M. Smith, pp. 101–32. Berlin: Springer-Verlag.

Steel, W. H. (1970). Fringe localization and visibility in classical and hologram interferometers. *Optica Acta*, **17**, 873–81.

Stetson, K. A. (1967*a*). Holography with total internally reflected light. *Applied Physics Letters*, **11**, 225–6.

Stetson, K. A. (1967*b*). Holographic fog penetration. *Journal of the Optical Society of America*, **57**, 1060–1.

Stetson, K. A. (1968*a*). Improved resolution and signal-to-noise ratios in total internal reflection holograms. *Applied Physics Letters*, **12**, 362–4.

Stetson, K. A. (1968*b*). Holographic surface contouring by limited depth of focus. *Applied Optics*, **7**, 987–9.

Stetson, K. A. (1969). A rigorous theory of the fringes of hologram interferometry. *Optik*, **29**, 386–400.

Stetson, K. A. (1970*a*). Moiré method for determining bending moments from hologram interferometry. *Optics & Laser Technology*, **2**, 80–4.

Stetson, K. A. (1970*b*). The argument of the fringe function in hologram interferometry of general deformations. *Optik*, **31**, 576–91.

Stetson, K. A. (1970*c*). Effects of beam modulation on fringe loci and localization in time-average hologram interferometry. *Journal of the Optical Society of America*, **60**, 1378–88.

Stetson, K. A. (1971). Hologram interferometry of nonsinusoidal vibrations analysed by density functions. *Journal of the Optical Society of America*, **61**, 1359–62.

Stetson, K. A. (1972*a*). Fringes of hologram interferometry for simple nonlinear oscillations. *Journal of the Optical Society of America*, **62**, 297–8.

Stetson, K. A. (1972*b*). Method of stationary phase for analysis of fringe functions in hologram interferometry. *Applied Optics*, **11**, 1725–31.

Stetson, K. A. (1974). Fringe interpretation for hologram interferometry of rigid body motions and homogeneous deformations. *Journal of the Optical Society of America*, **64**, 1–10.

Stetson, K. A. (1975*a*). Fringe vectors and observed-fringe vectors in hologram interferometry. *Applied Optics*, **14**, 272–3.

Stetson, K. A. (1975*b*). Homogeneous deformations: determination by fringe vectors in hologram interferometry. *Applied Optics*, **14**, 2256–9.

Stetson, K. A. (1976). Holographic strain analysis by fringe localization planes. *Journal of the Optical Society of America*, **66**, 627.

Stetson, K. A. (1978). The use of an image derotator in hologram interferometry and speckle photography of rotating objects. *Experimental Mechanics*, **18**, 67–73.

Stetson, K. A. (1979). Use of projection matrices in hologram interferometry. *Journal of the Optical Society of America*, **69**, 1705–10.

Stetson, K. A. & Powell, R. L. (1965). Interferometric hologram evaluation and real-time vibration analysis of diffuse objects. *Journal of the Optical Society of America*, **55**, 1694–5.

Stetson, K. A. & Taylor, P. A. (1971). The use of normal mode theory in holographic vibration analysis with application to an asymmetrical circular disk. *Journal of Physics E: Scientific Instruments*, **4**, 1009–15.

Stewart, W. C., Mezrich, R. S., Cosentino, L. S., Nagle, E. M., Wendt, F. S. & Lohman, R. D. (1973). An experimental read–write holographic memory, *RCA Review*, **34**, 3–44.

Stirn, B. A. (1975). Recording 360° holograms in the undergraduate laboratory. *American Journal of Physics*, **43**, 297–300.

Stroke, G. W. (1965). Lensless Fourier-transform method for optical holography. *Applied Physics Letters*, **6**, 201–3.

Stroke, G. W. (1966). White-light reconstruction of holographic images using transmission holograms recorded with conventionally focused images and in-line background. *Physics Letters*, **23**, 325–7.

Stroke, G. W., Brumm, D. & Funkhouser, A. (1965). Three-dimensional holography with 'lensless' Fourier-transform holograms and coarse P/N Polaroid film. *Journal of the Optical Society of America*, **55**, 1327–8.

Stroke, G. W. & Restrick, R. C. (1965). Holography with spatially non-coherent light. *Applied Physics Letters*, **7**, 229–30.

Stroke, G. W., Restrick, R., Funkhouser, A. & Brumm, D. (1965). Resolution-retrieving compensation of source effects by correlation reconstruction in high resolution holography. *Physics Letters*, **18**, 274–5.

Stroke, G. W. & Zech, R. G. (1966). White light reconstruction of color images from black and white volume holograms recorded on sheet film. *Applied Physics Letters*, **9**, 215–18.

Stroke, G. & Zech, R. G. (1967). A posteriori image-correcting 'deconvolution' by holographic Fourier-transform division. *Physics Letters*, **25A**, 89–90.

Su, F. & Gaylord, T. K. (1972). Calculation of arbitrary-order diffraction efficiencies of thick gratings with arbitrary grating shape. *Journal of the Optical Society of America*, **62**, 802–6.

Sugaya, T., Ishikawa, M., Hoshino, I. & Iwamoto, A. (1981). Holographic system for filing and retrieving patents. *Applied Optics*, **20**, 3104–8.

Suhara, T., Nishihara, H. & Koyama, J. (1975). The modulation transfer function in the hologram copying process. *Optics Communications*, **14**, 35–8.

Sutherlin, K. K., Lauer, J. P. & Olenick, R. W. (1974). Holoscan: a commercial holographic ROM. *Applied Optics*, **13**, 1345–54.

Suzuki, M., Saito, T. & Matsuoka, T. (1978). Multicolor rainbow hologram. *Kogaku*, **7**, 29–31.

Sweatt, W. C. (1977). Achromatic triplet using holographic optical elements. *Applied Optics*, **16**, 1390–1.

Sweeney, D. W. & Vest, C. M. (1973). Reconstruction of three-dimensional refractive index fields from multidirectional interferometric data. *Applied Optics*, **12**, 2649–64.

Tai, A., Yu, F. T. S. & Chen, H. (1979). Multislit one-step rainbow holographic interferometry. *Applied Optics*, **18**, 6–7.

Takai, N., Yamada, M. & Idogawa, T. (1976). Holographic interferometry using a reference wave with a sinusoidally modulated amplitude. *Optics & Laser Technology*, **8**, 21–3.

Tamura, P. N. (1977). Multicolor image from superposition of rainbow holograms. In *Clever Optics: Innovative Applications of Optics*, Proceedings of the SPIE, vol. 126, ed. N. Balasubramanian & J. C. Wyant, pp. 59–66. Redondo Beach: SPIE.

Tamura, P. N. (1978a). Pseudocolor encoding of holographic images using a single wavelength. *Applied Optics*, **17**, 2532–6.

Tamura, P. N. (1978b). One step rainbow holography with a field lens. *Applied Optics*, **17**, 3343.

Tanner, L. H. (1966). Some applications of holography in fluid mechanics. *Journal of Scientific Instruments*, **43**, 81–3.

Thaxter, J. B. & Kestigian, M. (1974). Unique properties of SBN and their use in a layered optical memory. *Applied Optics*, **13**, 913–24.

The Optical Society of America (1953). *The Science of Color*. New York: Thomas Y. Crowell.

Thinh, V. N. & Tanaka, S. (1973). Real time interferometry using thermoplastic hologram. *Japanese Journal of Applied Physics*, **12**, 1954–5.

Thompson, B. J. (1963). Fraunhofer diffraction patterns of opaque objects with coherent background. *Journal of the Optical Society of America*, **53**, 1350.

Thompson, B. J. (1974). Holographic particle sizing techniques. *Journal of Physics E: Scientific Instruments*, **7**, 781–8.

Thompson, B. J., Ward, J. H. & Zinky, W. R. (1967). Application of hologram techniques for particle size analysis. *Applied Optics*, **6**, 519–26.

Tischer, F. J. (1970). Analysis of ghost images in holography by the use of Chebyshev polynomials. *Applied Optics*, **9**, 1369–74.

Tomlinson, W. J. & Aumiller, G. D. (1975). Technique for measuring refractive index changes in photochromic materials. *Applied Optics*, **14**, 1100–4.

Tonin, R. & Bies, D. A. (1978). General theory of time-averaged holography for the study of three-dimensional vibrations at a single frequency. *Journal of the Optical Society of America*, **68**, 924–31.

Trolinger, J. D. (1975a). Particle field holography. *Optical Engineering*, **14**, 383–92.

Trolinger, J. D. (1975b). Flow visualization holography. *Optical Engineering*, **14**, 470–81.

Trukhmanova, T. D. & Denisyuk, G. V. (1977). Investigation of the applicability of locally manufactured fine-grain emulsions for obtaining relief structures. *Zhurnal Nauchnoi i Prikladnoi Fotografii i Kinematografii*, **32**, 178–81.

Tsujiuchi, J. (1963). Correction of optical images by compensation of aberrations and by spatial frequency filtering. In *Progress in Optics*, vol. 2, ed. E. Wolf, pp. 133–82. Amsterdam: North-Holland.

Tsukamoto, K., Ishii, A., Ishida, A., Sumi, M. & Uchida, N. (1974). Holographic information retrieval system. *Applied Optics*, **13**, 869–74.

Tsunoda, Y., Tatsumo, K. & Kataoka, K. (1976). Holographic video disk: an alternative approach to optical video disks. *Applied Optics*, **15**, 1398–403.

Tsuruta, T., Shiotake, N. & Itoh, Y. (1968). Hologram interferometry using two reference beams. *Japanese Journal of Applied Physics*, **7**, 1092–100.

Tsuruta, T., Shiotake, N., Tsujiuchi, J. & Matsuda, K. (1967). Holographic generation of contour map of diffusely reflecting surface by using immersion method. *Japanese Journal of Applied Physics*, **6**, 661–2.

Uchida, N. (1973). Calculation of diffraction efficiency in hologram gratings attenuated along the direction perpendicular to the grating vector. *Journal of the Optical Society of America*, **63**, 280–7.

Uozato, H. & Nagata, R. (1977). Holographic photoelasticity by using dual-hologram method. *Japanese Journal of Applied Physics*, **16**, 95–100.

Upatnieks, J. (1967). Improvement of two-dimensional image quality in coherent optical systems. *Applied Optics*, **6**, 1905–10.

Upatnieks, J. & Embach, J. T. (1980). 360-degree hologram displays. *Optical Engineering*, **19**, 696–704.

Upatnieks, J. & Leonard, C. (1969). Diffraction efficiency of bleached, photographically recorded interference patterns. *Applied Optics*, **8**, 85–9.

Upatnieks, J. & Leonard, C. (1970). Efficiency and image contrast of dielectric holograms. *Journal of the Optical Society of America*, **60**, 297–305.

Upatnieks, J., Marks, J. & Federowicz, R. (1966). Color holograms for white light reconstruction. *Applied Physics Letters*, **8**, 286–7.

Upatnieks, J., Vander Lugt, A. & Leith, E. (1966). Correction of lens aberrations by means of holograms. *Applied Optics*, **5**, 589–93.

Urbach, J. C. (1977). Thermoplastic hologram recording. In *Holographic Recording Materials*, Topics in Applied Physics, vol. 20, ed. H. M. Smith, pp. 161–207. Berlin: Springer-Verlag.

Urbach, J. C. & Meier, R. W. (1966). Thermoplastic xerographic holography. *Applied Optics*, **5**, 666–7.

Vagin, L. N., Nazarova, L. G., Arseneva, T. M. & Vanin, V. A. (1975). Holographic miniaturization of scientific and technical documents. *Optics & Spectroscopy*, **38**, 571–3.

Vagin, L. N. & Shtan'ko, A. E. (1974). Copying holograms by stamping on a thermoplastic. *Optics & Spectroscopy*, **36**, 597–8.

van Deelen, W. & Nisenson, P. (1969). Mirror blank testing by real time holographic interferometry. *Applied Optics*, **8**, 951–5.

Vander Lugt, A. (1964). Signal detection by complex spatial filtering. *IEEE Transactions on Information Theory*, IT-10, 139–45.

Vander Lugt, A. (1973). Design relationships for holographic memories. *Applied Optics*, **12**, 1675–85.

Vander Lugt, A., Rotz, F. B. & Klooster Jr, A. (1965). Character reading by optical spatial filtering. In *Optical & Electro-Optical Information Processing*, ed. J. T. Tippett, D. A. Berkowitz, L. C. Clapp, C. J. Koester & A. Vanderburgh Jr, pp. 125–41. Cambridge: Massachusetts Institute of Technology Press.

van Heerden, P. J. (1963a). A new optical method of storing and retrieving information. *Applied Optics*, **2**, 387–92.

van Heerden, P. J. (1963b). Theory of optical information storage in solids. *Applied Optics*, **2**, 393–400.

Vanin, V. A. (1978). Hologram copying. *Soviet Journal of Quantum Electronics*, **8**, 809–18.

Vanin, V. A. (1979). Influence of the polarization of the object and reference waves on the hologram quality. *Soviet Journal of Quantum Electronics*, **9**, 774–6.

van Ligten, R. (1966). Influence of photographic film on wavefront reconstruction. II. 'Cylindrical' wavefronts. *Journal of the Optical Society of America*, **56**, 1009–14.

van Ligten, R. F. & Osterberg, H. (1966). Holographic microscopy. *Nature*, **211**, 282–3.

van Renesse, R. L. (1980). Scattering properties of fine-grained bleached emulsions. *Photographic Science & Engineering*, **24**, 114–19.

Velzel, C. H. F. (1973). Non-linear holographic image formation. *Optica Acta*, **20**, 585–606.

Vest, C. M. (1979). *Holographic Interferometry*. New York: John Wiley.

Vest, C. M. (1981). *Holographic NDE: Status & Future*. National Bureau of Standards Report NBS-GCR-81-318. Springfield: National Technical Information Service.

Vikram, C. S. (1974a). Quadruple-exposure holographic interferometry for analysis of superposition of ramp motion and sinusoidal vibration. *Optics Communications*, **10**, 290–1.

Vikram, C. S. (1974b). Stroboscopic holographic interferometry of vibration simultaneously in two sinusoidal modes. *Optics Communications*, **11**, 360–4.

Vikram, C. S. (1975). Holographic interferometry of superposition of two motions. *Optik*, **43**, 65–70.

Vikram, C. S. (1976). Holographic interferometry of superposition of motions with different time functions. *Optik*, **45**, 55–64.

Vikram, C. S. (1977). Mechanical vibrations: mapping their phase with hologram interferometry. *Applied Optics*, **16**, 1140–1.

Vilkomerson, D. H. R. & Bostwick, D. (1967). Some effects of emulsion shrinkage on a hologram's image space. *Applied Optics*, **6**, 1270–2.

Vlasov, N. G., Ryabova, R. V. & Semenov, S. P. (1977). Leith holograms reconstructed in white light. *Zhurnal Nauchnoi i Prikladnoi Fotografii i Kinematografii*, **22**, 384–5.

von Bally, G. ed. (1979). *Holography in Medicine and Biology*. Berlin: Springer-Verlag.

Walles, S. (1969). Visibility and localization of fringes in holographic interferometry of diffusely reflecting surfaces. *Arkiv för Fysik*, **40**, 299–403.

Walkup, J. F. (1980). Space-variant coherent optical processing. *Optical Engineering*, **19**, 339–46.

Waters, J. P. (1968). Three-dimensional Fourier transform method for synthesizing binary holograms. *Journal of the Optical Society of America*, **58**, 1284–8.

Waters, J. P. (1972). Object motion compensation by speckle reference beam interferometry. *Applied Optics*, **11**, 630–6.

Watrasiewicz, B. M. & Spicer, P. (1968). Vibration analysis by stroboscopic holography. *Nature*, **217**, 1142–3.

Weaver, C. S. & Goodman, J. W. (1966). A technique for optically convolving two functions. *Applied Optics*, **5**, 1248–9.

Webster, J. M., Tozer, B. A. & Davis, C. R. (1979). Holography of large volumes using holographic scatter plates. *Optics & Laser Technology*, **11**, 157–9.

Weingärtner, I. & Rosenbruch, K. J. (1980a). Incoherent polychromatic imaging with holographic optical elements and systems. *Optik*, **57**, 103–22.

Weingärtner, I. & Rosenbruch, K. J. (1980b). Incoherent polychromatic imaging with holographic optical elements and systems. II. *Optik*, **57**, 161–71.

Weingärtner, I. & Rosenbruch, K. J. (1982). Chromatic correction of two- and three-element holographic imaging systems. *Optica Acta*, **29**, 519–29.

Welford, W. T. (1973). Aplanatic hologram lenses on spherical surfaces. *Optics Communications*, **9**, 268–9.

Wilson, A. D. (1970). Characteristic functions for time-average holography. *Journal of the Optical Society of America*, **60**, 1068–71.

Wilson, A. D. (1971). Computed time-average holographic interferometric fringes of a circular plate vibrating simultaneously in two rationally or irrationally related modes. *Journal of the Optical Society of America*, **61**, 924–9.

Wilson, A. D. & Strope, D. H. (1970). Time-average holographic interferometry of a circular plate vibrating simultaneously in two rationally related modes. *Journal of the Optical Society of America*, **60**, 1162–5.

Windischbauer, G., Keck, F. G., Cabaj, A., Langschwert, H., Ranninger, G. & Tomiser, J. (1973). Polarization holography – a critical review. *Optik*, **37**, 385–90.

Wolfke, M. (1920). Über der Möglichkeit der optischen Abbildung vom Molekulargittern. *Physikalische Zeitschrift*, **21**, 495–7.

Worthington Jr, H. R. (1966). Production of holograms with incoherent illumination. *Journal of the Optical Society of America*, **56**, 1397–8.

Woznicki, J. (1980). Geometry of recording and colour sensitivity for evanescent wave holography using a Gaussian beam. *Applied Optics*, **19**, 631–7.

Wuerker, R. F. & Heflinger, L. O. (1970). Pulsed laser holography. In *The Engineering Uses of Holography*, ed. E. R. Robertson & J. M. Harvey, pp. 99–114. Cambridge: University Press.

Wüthrich, A. & Lukosz, W. (1975). Holography with evanescent waves. II. Measurements of the diffraction efficiencies. *Optik*, **42**, 315–34.

Wüthrich, A. & Lukosz, W. (1980). Holography with guided optical waves. I. Experimental techniques and results. *Applied Physics*, **21**, 55–64.

Wyant, J. C. (1977). Image blur for rainbow holograms. *Optics Letters*, **1**, 130–2.

Wyant, J. C. & Bennett, V. P. (1972). Using computer-generated holograms to test aspheric wavefronts. *Applied Optics*, **11**, 2833–9.

Wyant, J. C. & O'Neill, P. K. (1974). Computer generated hologram: null lens test of aspheric wavefronts. *Applied Optics*, **13**, 2762–5.

Yan-Song, C., Yu-Tang, W. & Bi-Zhen, D. (1978). A new method of colour holography. *Acta Physica Sinica*, **27**, 723–8.

Yaroslavskii, L. P. & Merzlyakov, N. S. (1980). *Methods of Digital Holography*. New York: Consultants Bureau, Plenum Publishing Co.

Yatagai, T. & Saito, H. (1978). Interferometer testing with computer-generated holograms: aberration balancing method and error analysis. *Applied Optics*, **17**, 558–65.

Yatagai, T. & Saito, H. (1979). Dual computer-generated holograms for testing aspherical surfaces. *Optica Acta*, **26**, 985–93.

Young, M. & Hicks, A. (1974). Holographic ruby laser with long coherence and precise timing. *Applied Optics*, **13**, 2486–8.

Yu, F. T. S. & Chen, H. (1978). Rainbow holographic interferometry. *Optics Communications*, **25**, 173–5.

Yu, F. T. S., Ruterbusch, P. H. & Zhuang, S. L. (1980). High-resolution rainbow holographic process. *Optics Letters*, **5**, 443–5.

Yu, F. T. S., Tai, A. & Chen, H. (1978). Archival storage of color films by rainbow holographic technique. *Optics Communications*, **27**, 307–10.

Yu, F. T. S., Tai, A. & Chen, H. (1979). Multiwavelength rainbow holographic interferometry. *Applied Optics*, **18**, 212–18.

Yu, F. T. S., Tai, A. M. & Chen, H. (1980). One-step rainbow holography: recent development and application. *Optical Engineering*, **19**, 666–78.

Yu, F. T. S. & Wang, E. Y. (1973). Speckle reduction in holography by means of random spatial sampling. *Applied Optics*, **12**, 1656–9.

Zaidel', A. N., Ostrovskaya, G. V. & Ostrovskii, Yu. I. (1969). Plasma diagnostics by holography. *Soviet Physics – Technical Physics*, **13**, 1153–64.

Zambuto, M. & Lurie, M. (1970). Holographic measurement of general forms of motion. *Applied Optics*, **9**, 2066–72.

Zambuto, M. H. & Fischer, W. K. (1973). Shifted reference holographic interferometry. *Applied Optics*, **12**, 1651–5.

Zelenka, J. S. & Varner, J. R. (1968). New method for generating depth contours holographically. *Applied Optics*, **7**, 2107–10.

Zelenka, J. S. & Varner, J. R. (1969). Multiple-index holographic contouring. *Applied Optics*, **8**, 1431–4.

Zemtsova, E. G. & Lyakhovskaya, L. V. (1976). A study of a method of copying three-dimensional holograms. *Soviet Journal of Optical Technology*, **43**, 744–6.

Zetsche, C. (1982). Simplified realization of the holographic inverse filter: a new method. *Applied Optics*, **21**, 1077–9.

Author index

Subject index

Abel transform, 216
aberrations, 29–32, 186, 192
  classification of, 30
  correction of, 7, 177–9, 192, 193
  of holographic optical elements, 192
  problems in microscopy, 174
absorption holograms, *see* amplitude
    holograms
achromatic images, 9, 143–5, 192
  by dispersion compensation, 143
  from holographic stereograms, 156–8
  with image holograms, 21, 34, 143
  with rainbow holograms, 9, 144–5
  with white-light recording, 169
aerosols, *see* particle-size analysis
aliasing, 6, 255
amplitude holograms, 41, 78
  thin, 11, 41–3
  volume transmission, 49–51
  volume reflection, 52–6
amplitude transmittance, 12
  and H & D curve, 271
  complex, 78
  vs. exposure curve, 14, 78, 79
  vs. log-exposure curve, 80–1
analytic signal, 261
anamorphic image, 32
angular magnification, 26
argon-ion laser, 34, 68, 70, 71, 73, 101
  for colour holography, 131
  for contouring, 247
  mode structure, 72
  output wavelengths, 73
  use with photoresists, 106, 189
associative storage, 194–6
autocorrelation, 254

beam couplers, 61
beam ratio, 66, 69
beam splitters, 61, 66–7, 77
  holographic, 61, 193
  polarizing prism, 67
$Bi_{12}GeO_{20}$, 113, 217
binary holograms, 147–51, 158

$Bi_{12}SiO_{20}$, 113–15
  characteristics, 89, 114
  for contouring, 247
  for phase-conjugate imaging, 179
  for vibration analysis, 236
bleach techniques, 96–9
  conventional, 96
  light stability, 97–9, 100
  noise, 96–7
  reversal, 97, 99
  silver-halide sensitized gelatin, 104–6
  tanning, 96, 170
Borrmann effect, 60
Bragg condition, 43, 45, 46, 47, 48, 49, 51,
    52, 53, 55, 56, 58, 60, 62, 113, 203
  effect of emulsion shrinkage, 62
  effect of change in permittivity, 58
  in three dimensions, 203
  in multiply-exposed holograms, 60
Bragg x-ray microscope, 2

carrier frequency, 6, 14
  minimum for no overlap, 16
character recognition, 7, 194, 196–9
characteristic function, 233, 234, 237, 239,
    240, 241
  generalized, 241
  method of stationary phase, 241
  probabilistic interpretation, 240
chromaticity diagram, 131, 132
cinematography, 130
coded reference wave
  for multicolour imaging, 133
  for multiplexing images, 199
  for polarization recording, 163
  for space-variant image processing, 203
coherence, 262–4
  of laser light, 69–72
  length, 69, 71, 73, 264
  requirements for holography, 69
  spatial, 69, 263
  temporal, 69, 70, 263–4
colour holography, 8, 131–45
  cross-talk images, 133
  image luminance, 135–6, 139

313

emulsion shrinkage (*continued*)
   methods of control, 93, 97, 135
   production of relief images, 92–3
erasable recording materials, *see*
   photochromics, photothermoplastics,
   photorefractive crystals
etalon, 71–2, 73, 75
evanescent-wave holography, 181–5
   reconstruction with white light, 183–4
exposure, 270
   control, 69
exposure characteristics
   ideal material, 78
   available materials, 89–90

far-field condition, 22, 175, 258
far-field hologram, 22, 175
fast Fourier transform, 146
ferroelectric crystals, *see* photorefractive
   crystals
four-wave mixing, *see* phase-conjugate
   imaging
Fourier holograms, 17–19
   for cinematography, 130
   for multiple imaging, 185–6
   resolution of image, 81
   formation of stationary image, 19
   see also lensless Fourier holograms
Fourier transform, 252–3
   by a lens, 259–60
   discrete, 146, 150, 255–6
   fast, 146
   of a sampled function, 255
Fraunhofer hologram, 21–2, 175–6
frequency-translated holography, 238–9
Fresnel–Kirchhoff integral, 258
fringe control, 212
fringe localization, *see* localization of
   fringes
fringe-locus function, 226
fringe-vector method, 226
frost, 111

Gabor hologram, *see* in-line hologram
gamma, 271
gas lasers, 65, 69, 72–3
   wavelengths, 73
Gaussian beam, 67–8
ghost imaging, 196
grating vector, 44

H & D curve, 270–1
   and |t| vs. E curve, 271
head-up displays, 193
helium–cadmium laser, 73
   for colour holography, 131
   use with photoresists, 106

helium–neon laser, 71, 73
   for colour holography, 131
   mode structure, 73
heterodyne interferometry, 228–9
high-resolution imaging, 177–85
   with evanescent-wave holograms, 184–5
historical development, 2–6
holodiagram, 69, 76, 223–5
holographic gratings, 187–9
holographic interferometry, 7, 207–51
   electronic techniques, 228–31
   evaluation of surface displacements, 217–25
   in an industrial environment, 214–15
   in fluid flow, 207, 215
   in medical research, 207
   in plasma diagnostics, 207, 217
   object-motion compensation, 215
   of phase objects, 215–17
   of rotating objects, 214
   phase-conjugate interferometry, 217
   phase-difference amplification, 217
   sensitivity vector, 219, 224, 228, 232
   two-hologram technique, 212
   with pulsed lasers, 214
   see also contouring, double-exposure
      interferometry, holodiagram,
      localization of fringes, photoelasticity,
      real-time interferometry, sandwich
      holograms, strain analysis, vibration
      analysis
holographic optical elements, 7, 191–3
holographic scanners, 189–91
holographic stereograms, 9, 119–20
   for contouring, 251
   from computer-generated views, 156–8
   image blur, 129–30
   image distortion, 129
   of living subjects, 119
   reconstruction with white light, 9, 128–9
holographic subtraction, 211, 240
hyperstereoscopic distortion, 119

image aberrrations, 29–32
image blur
   and source size, 33
   and spectral-bandwidth, 33–4
   in achromatic images, 144
   in multicolour images, 143
   in pseudocolour images, 142, 143
   with holographic stereograms, 129–30
   with image holograms, 34, 37
   with rainbow holograms, 125–7
image holograms, 21
   for contouring, 251
   for holographic interferometry, 215
   increased image luminance, 21, 36–7,
      135–6, 143